Regenerative Medicine, Artificial Cells and Nanomedicine – Vol. 5

THE LUNG

Developmental Morphogenesis,
Mechanobiology, and Stem Cells

Regenerative Medicine, Artificial Cells and Nanomedicine

ISSN: 2010-2836

Series Editor: Thomas Ming Swi Chang

Published

Regenerative Medicine, Artificial Cells and Nanomedicine – Vol. 5

THE LUNG

Developmental Morphogenesis, Mechanobiology, and Stem Cells

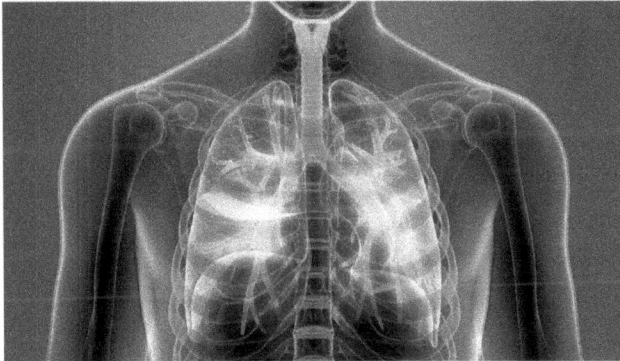

Ahmed El-Hashash
University of Edinburgh | Zhejiang

Eiman Abdel Meguid
Queen's University Belfast

World Scientific

NEW JERSEY · LONDON · SINGAPORE · BEIJING · SHANGHAI · HONG KONG · TAIPEI · CHENNAI

Published by

World Scientific Publishing Co. Pte. Ltd.

5 Toh Tuck Link, Singapore 596224

USA office: 27 Warren Street, Suite 401-402, Hackensack, NJ 07601

UK office: 57 Shelton Street, Covent Garden, London WC2H 9HE

Library of Congress Cataloging-in-Publication Data

Names: El-Hashash, Ahmed, author. | Meguid, Eiman Abdel, author.
Title: The lung : developmental morphogenesis, mechanobiology, and stem cells / by
    Ahmed El-Hashash (The Zhejiang University-University of Edinburgh Institute, China) and
    Eiman Abdel Meguid (Queen's University Belfast, UK).
Other titles: Regenerative medicine, artificial cells and nanomedicine ; v. 5.
Description: New Jersey : World Scientific, 2019. | Series: Regenerative medicine,
    artificial cells and nanomedicine ; volume 5 | Includes bibliographical references.
Identifiers: LCCN 2018044016 | ISBN 9789813277069 (hc : alk. paper)
Subjects: | MESH: Lung--growth & development | Lung--physiology | Stem Cells | Cell Biology
Classification: LCC QP121 | NLM WF 600 | DDC 612.2/4--dc23
LC record available at https://lccn.loc.gov/2018044016

British Library Cataloguing-in-Publication Data
A catalogue record for this book is available from the British Library.

For any available supplementary material, please visit
https://www.worldscientific.com/worldscibooks/10.1142/11178#t=suppl

Desk Editors: Herbert Moses/Jennifer Brough/Shi Ying Koe

Typeset by Stallion Press
Email: enquiries@stallionpress.com

This book is dedicated to my Parents, Kids HOOR & NOOR, Wife and new baby LIEN

Ahmed El-Hashash, PhD

To my mother, Heba Nasr and my sons, Hossam Amr and Omar Amr, who made my life worthwhile. Thanks for your love and support. This book is dedicated to you.

Eiman Abdel Meguid, PhD

Foreword by Alexzander A. A. Asea

University of Texas, MD Anderson Cancer Center,
Houston, USA

It is with great pleasure that I pen this foreword to the *Lung: Developmental Morphogenesis, Mechanobiology, and Stem Cells*. The fields of lung developmental and stem cell biology and mechanobiology are moving extremely rapidly as the concept and potential practical applications have moved from theoretical concepts into human clinical trials, often with outstanding results, and has thus entered the mainstream. Despite this worldwide intensity and diversity of endeavor, there remain a smaller number of book volumes that are focused almost entirely on the lung developmental and stem cell biology and mechanobiology, and this volume is an example of this kind of book that also brings the novel scientific findings of most of the leaders of this research field together. Much of this book, therefore, concentrates on the fundamental lung developmental and cell biology and its stem cell from which the promising clinical applications will arise.

Better understanding of the lung developmental morphogenesis and mechanobiology and lung stem cell biology, which are the major goals of this book, will help in developing novel strategies for lung repair and regeneration after injury. In addition, the recent progress in our understanding of different aspects of lung stem cell biology and behavior, together with other types of stem cells and their therapeutic applications point to the impressive possibilities of patient-specific ad *hominem* treatment. This book provides a timely, up-to-date state-of-the-art reference of lung developmental morphogenesis and mechanobiology as well as lung stem cell biology.

Foreword by Junfeng Ji

Chairman, Dr. Li Dak Sum & Yip Yio Chin
Centre of Stem Cell and Regenerative Medicine,
School of Medicine, Zhejiang University, China

I am delighted to compose this foreword to the *Lung: Developmental Morphogenesis, Mechanobiology, and Stem Cells*. Lung developmental mechanobiology is a rapidly growing research field. This book covers recent progresses in the field of lung and stem cell biology and mechanobiology, and much of it properly concentrates on the fundamental of lung development, morphogenesis and mechanobiology.

Recent progress has been made toward better understanding many aspects of the lung developmental morphogenesis and mechanobiology. These are pertinent areas that this book covers. In addition, this book provides a timely, up-to-date and state-of-the-art reference in the lung stem cell biology and behavior, from which the highly potential applications may arise.

Regenerative biology and medicine are rapidly growing branches of translational research in both molecular biology and tissue engineering that deal with the "process of replacing, engineering or regenerating human cells, tissues or organs to restore or establish normal function". Regenerative biology holds many promises for the engineering of a damaged tissue or organ such as the lung. Recent discovers in both regenerative medicine and stem cell biology can be applied in the diseased or defected lung, but still have many challenges. Indeed, many accumulated recent research studies have focused on how we can harness the endogenous stem cell power as a source for tissue/organ regeneration in the lung, for example, which is well covered in this book.

Preface

The development of all humans begins after the union of male and female gametes or germ cells during fertilization or conception. The fertilized egg or zygote is a large diploid cell that is the beginning, or primordium, of a human being. This fertilized egg undergoes round after round of both highly organized and tightly controlled cell divisions until it comprises many billions of stem and lineage-specific cells that have self-renewal and self-repairing capabilities and form the human body. These processes are studied by a branch of science called developmental biology that explores how organisms develop and progress. As a lung developmental and stem cell biologist, who has investigated the mechanisms of organogenesis in a wide variety of tissues and organs such as the lung, placenta, kidney, and neural crest cells, it becomes clear to me that if we can understand these normal and fundamental mechanisms of developmental biology and mechanobiology, then correcting abnormalities caused by various lung diseases and congenital defects, repairing the injured tissues and even generation of functional whole organs from stem cells should be theoretically achievable.

The stem cell field has grown very rapidly over the past decade, and continues to be one of the most exciting aspects of biomedical research. Stem cell research can be traced back to more than 20 years when scientists first isolated embryonic stem cells from mouse blastocysts, and a research article announcing the discovery of human embryonic stem cells emerged in 1998. Stem cell research is a fast-growing field that has rapidly expanded as new research and experience broaden our knowledge about different aspects of stem cell biology and applications. In the past decade, the stem cell field has grown very rapidly, and continues to be one of the most exciting aspects of biomedical research.

Both embryonic and adult stem cells are currently a remarkably fast-growing field of research, with an astonishing annual growth rate of 77% since 2008. The volume of research output, and thus publication, has therefore increased significantly in all areas of stem cell research. By now, the first functioning whole organ, thymus, has been generated in the laboratory, and the first in vitro fertilized human baby girl has children of her own. Research is currently underway in different laboratories worldwide to generate other functioning whole organs such as the intestine, kidney and other human body organs.

Embryonic stem cells (ESCs) were isolated from mouse blastocysts by scientists in 1981, while human ESCs were first reported in 1998. Currently, adult-derived stem cells (ASCs) are also a favorite subject of intensive research investigations. Recently, ESCs have become almost routine in the face of more advances in the medical field. More recent advances show the possibility to turn fully differentiated cells back into a more embryonic-like state of induced pluripotency. This occurs using as few as four factors and represents a major scientific discovery. Moreover, it has been recently shown that several classes of stem-like cells, which originate from different mesenchymal compartments of the body such as the amnion, marrow, amniotic fluid and adipose, exert promising therapeutic effects in some inflammatory and fibrotic diseases. In addition, neural stem cells can be programmed to selectively travel and attack inaccessible brain tumors. Furthermore, the recent identification of endophenotypes, or latent risk factors, for certain types of aggressive cancers may eventually lead to designing novel strategies for cancer treatments. Together, these recent discoveries could identify the next generation of treatments emerging from our scientific discoveries.

Scientists worldwide are applying new stem cell discoveries to the betterment of human diseases, bringing forth much hope for better human life. The branch of translational research in tissue engineering and molecular biology which takes advantage of rapid progress in our understanding of stem cell biology during development and adulthood is called regenerative medicine. The hope for cure for different diseases has prompted different countries worldwide to invest in stem cell research and regenerative medicine.

The United States of America, for example, plays a critical role in stem cell research, similar to a lot of other countries in the world. Many countries in Europe and Australia; Japan, China and other Asian countries; in addition to Canada and Brazil, have leading centers for stem cell research and regenerative medicine. These research centers have significantly expanded the scope of stem cell research and their applications in the treatment of different human diseases.

New insights have been gained on the identification and characterization of endogenous-tissue-specific stem and progenitor cells in the lung over the last few years. The exploration of endogenous lung stem and progenitor cells holds promise for advancing our understanding of the biology of lung repair and regeneration mechanisms after injury. This will also help in the future use of stem cell therapies for the development of regenerative medicine approaches for the treatment of different lung diseases.

This book brings together several topics that are related to lung developmental morphogenesis, mechanobiology as well as lung stem cell biology, and development and stem cell applications in lung repair and regeneration. We discuss recent advances in lung developmental morphogenesis and mechanobiology as well as the identification and characterization of the main types of lung cell populations. In addition, we describe recent research progresses and accumulated information regarding the behavior, development and function of various lung stem and progenitor cells, and factors that control their repair and regeneration after injury as well as the molecular and cellular mechanisms that regulate lung stem and progenitor cell behavior during development, repair and regeneration. Moreover, we describe recent advances in lung stem cell polarity and mode of divisions.

This book contains a review of a global collection of recent monograph essays from a wide range of research scientists who are investigating different aspects of lung developmental morphogenesis and mechanobiology and stem cells at various research institutes and in various countries. It describes and discusses exciting progresses in basic stem cell behavior and biology and regenerative medicine, including the potential applications of stem cells in lung repair and regeneration as well as in some lung diseases. In addition, this book describes advances in our understanding of stem cell-related diseases in the lung and stem cell-based therapies of different lung diseases as well as major progresses and challenges in lung regenerative medicine.

Although we cannot hope to be comprehensive in the coverage of lung developmental morphogenesis, mechanobiology and stem cells, our main goal in compiling this book was to bring together a selection of the current progress in understanding the developmental, stem cell and mechanobiology of the lung, as well as the potential applications of stem cells in lung repair and regeneration after injury or diseases. In preparing this book, we aimed at making it accessible not only to those working in lung cellular, developmental and mechanobiology fields but also to non-experts with a broad interest in lung biology, stem cells and regenerative biology. Our hope is that this book will be of value to all concerned with lung developmental, stem cell and mechanobiology.

About the Authors

Ahmed El-Hashash has completed his PhD from Manchester University, UK. He is a fellow of the California Institute of Regenerative Medicine (CIRM) and New York University Medical School (MSSM), USA. Professor Ahmed El-Hashash worked as a Senior Biomedical Research Scientist at the Mount Sinai School of Medicine of New York University and Children's Hospital Los Angeles. He was an Assistant Professor and the Principal Investigator of Stem Cell and Regenerative Medicine at the Keck School of Medicine and Ostrow School of Dentistry of The University of Southern California, USA. He has joined the University of Edinburgh, Edinburgh — Zhejiang International Campus, (ZJU) as the Tenure Track Associate Professor and Senior Principal Investigator of Biomedicine, Stem Cell and Regenerative Medicine. He is also the Adjunct Professor at the School of Basic Medical Science and School of Medicine, Zhejiang University. Hashash has several breakthrough discoveries in genes/enzymes that control stem cell behavior and regenerative medicine. He has published more than 34 papers and abstracts in reputed international journals and is serving as an editorial board member of repute. Professor El-Hashash acts as a discussion leader at the prestigious Gordon Research Seminar/Conference in USA, and a Peer Reviewer/International Extramural Review for The Medical Research Council (MRC) grant applications, London, UK. He has been invited to speak at several international conferences in USA, Spain, Greece, Egypt, and China. He is the editor or author of several books on stem cell and regenerative medicine.

Eiman Abdel Meguid, MBChB, PhD, P.G.C.H.E.T, F.H.E.A., is a Senior Lecturer at the Centre for Biomedical Sciences Education, School of Medicine, Dentistry and Biomedical Sciences, Queen's University Belfast, UK. Dr Abdel Meguid obtained her bachelor degree in Medicine and Surgery and her PhD in Anatomy and Embryology from Alexandria University. She completed her PhD thesis at Eberhard-Karls Universität Tübingen, Germany. Dr Abdel Meguid is an innovative educator and researcher. She has published multiple scientific works in leading journals, and she is a reviewer and a member of the editorial board of a number of journals. She taught gross anatomy to medical, dental, human biology students and students enrolled in MSc in clinical anatomy. Her research interests are in the areas of stem cells, lung development, anatomical pedagogy, teaching strategies, and the integration of novel technologies to enhance learning. She is a member of the Risk Management Group Committee and member of the Court of Examiners for the Royal College of Physicians and Surgeons, Glasgow. She is also a member of the Career Development Committee of the American Association of Clinical Anatomists.

Acknowledgments

The authors would like to thank the following persons:

Esam I Agamy, PhD
Professor, College of Medicine,
University of Sharjah, Sharjah,
UAE

Wadah AlHassan, BSc
California State Polytechnic University, Pomona,
3801 West Temple Avenue Pomona, California 91768,
USA

Alexzander A. A. Asea, PhD
Visiting Professor and Consultant Immunologist,
Center for Radiation Oncology Research,
Department of Experimental Radiation Oncology,
The University of Texas MD Anderson Cancer Center,
1515 Holcombe Boulevard,
Houston, Texas 77030,
USA

Mohamed Berika, PhD
Assistant Professor of Anatomy,
Anatomy Department, Faculty of Medicine,
Mansoura University, Mansoura, Egypt
and Rehabilitation Science Department,
College of Applied Medical Sciences,

King Saud University, Rehyad,
Kingdom of Saudi Arabia

Karen Ek, BSc
California State University San Bernardino,
5500 University Pkwy, San Bernardino,
California 92407,
USA

Magdy Elhefnawy, PhD
President, Gharbia Medical Syndicate,
Tanta University Medical School, Egypt

Haifen Huang, BSc
California State Polytechnic University, Pomona,
3801 West Temple Avenue,
Pomona, California 91768,
USA

Junfeng Ji, PhD
Professor of Stem Cells and Regenerative Medicine,
Chairman, Dr. Li Dak Sum & Yip Yio Chin Centre
for Stem Cell and Regenerative Medicine, School of Medicine,
Zhejiang University,
866 Yuhangtang Road, Hangzhou,
Zhejiang 310058, China

John Ku, BSc
California State Polytechnic University, Pomona,
3801 West Temple Avenue,
Pomona, California 91768,
USA

Linrong Lu, PhD
Professor of Immunology,
School of Basic Medical Sciences,
Zhejiang University,

866 Yuhangtang Road, Hangzhou,
Zhejiang 310058, China

Karol Lu, BSc
University of Southern California,
University Park, Los Angeles, California 90089,
USA

Gamal Madkour, PhD
Professor, Tanta University School of Science,
Tanta University, Egypt

Moustafa Mahmoud, PhD
Professor, Tanta University School of Medicine,
Tanta University, Egypt

Contents

Introduction

The respiratory system carries out its functions, including gas exchange, a few minutes after birth in newborn babies. The respiratory tract develops early in the embryonic life. It is divided into an upper respiratory tract, which consist of the nasal cavity and the pharynx, and a lower respiratory tract, which consist of the larynx, trachea, bronchi and the lungs. The lung possesses an extensively huge surface area that is almost 50–100 m² in humans, and its function is to exchange carbon dioxide and oxygen across a very thin membrane.

The main function of the lung is to exchange gas. The lung goes through five distinct histological phases of development. The two main respiratory cell types are the squamous alveolar type 1 and alveolar type 2 (surfactant secreting cells). They both arise from the same progenitor cells. The third main cell type is the macrophages (dust cells) which arise from blood monocyte cells.

Lung growth proceeds through gestation. There is progressive branching of the bronchioles and, finally, development of alveoli during the last trimester. The lung possesses many branched ducts that conduct air to and from the alveolar gas exchange surface in a way that increases the gas exchange surface between the blood and the atmosphere. Human lungs weigh approximately 1.3 kg.

The respiratory system is characterized by huge cellular diversity with multiple mesenchymal and epithelial lineages. These components are organized as a complex system of branched tubules in close relation to vascular and lymphatic conduits, which are in connection with a vast network of gas exchanging units, termed as the alveoli. This complex structure is developed sequentially by early branching of the epithelial tube, followed by septation of the air sacs. In conjunction with this step, the pulmonary vessels develop to help in transporting of gases to and from the alveoli. Development of this system is completed just before birth. Lung growth also continues after delivery as alveoli continue to increase in number.

The achievement of functionality and lung maturation is under the control of several hormones. At the end of the 6th month of gestation in humans, alveolar cells type 2 are formed and start to produce surfactant, and therefore premature babies have difficulties because they have insufficient surfactant. Surfactant is formed by phospholipids, and they help minimize the surface tension of the alveoli and prevent alveolar collapse during expiration. It develops in the last trimester of pregnancy, and reaches maturity by the 36th weeks.

Chapter 1

Lung Pattern Formation and Development

Abstract

The major function of the lung in different organisms is to perform an efficient exchange of gas with the atmosphere. The thickness of the surface of gas diffusion of the mature lung is 1 micron in humans. However, this produces a surface area of 70 square meters, which is equivalent in size to a modern tennis court or the wing surface of a small aircraft. The lung has a complex organization within the chest as a honeycomb-like structure that comprises a network of extensively branched ducts that function to conduct air to and from the alveolar gas exchange surface. This occurs in a configuration that remarkably increases the surface that facilitates gas exchange between blood and air, while enabling maximally efficient packing of this surface within the chest cavity. This complex structure of the lung is developed sequentially by early branching of the epithelial tube and later on by the process of the septation of terminal air sacs. In addition, the development of pulmonary vasculature that occurs in conjunction with epithelial branching morphogenesis acts to facilitate the transport of respiratory gases to and from the developing alveolar surface. In conjunction with these developmental processes, the development of airway smooth muscle (ASM) takes place during early lung morphogenesis, and its contraction may function to regulate the growth of the lung. Any perturbation of these tightly regulated developmental processes can lead to the formation of abnormal lung structure, gas exchange deficiency and/or respiratory failure. Such disruption of normal lung growth and development is clinically exemplified in many cases, such as premature human delivery, bronchopulmonary dysplasia, or congenital lung

1

defects or disorders. This chapter will describe different phases of lung development, genetic control of the formation pattern of early lung anlagen, distal airway branching morphogenesis, and the alveolar septum formation and its regulatory factors and molecular mechanisms as well as the development of various lung-specific cell types.

Keywords: Lung; developmental phases; pseudoglandular; canalicular; saccular; alveolar; pattern formation; branching morphogenesis; lung analgen.

1.1 Introduction

Throughout nature, branched networks are ubiquitous and particularly exist in various tissue types that require large surface area within a restricted volume. A wide range of tissues with a branched architecture, including the lung, kidney, vasculature, mammary gland and nervous system, function to exchange gases, fluids and information throughout the body of an organism. The generation of different branched tissue types requires a proper control of various biological processes, such as the specification, initiation and elongation of the branch site. Despite their complexity, the branching events in different organ types normally require the coordination of many cells to build a network of tissues for material exchange (Spurlin and Nelson, 2017).

The development of the respiratory system occurs through a series of steps that begin with the division of the common foregut tube into the respiratory endoderm anteriorly and the esophagus posteriorly. The respiratory tract undergoes excessive branching to form the proximal conducting airways, followed by distal septation that is able to generate the gas exchange units, called the alveoli. In addition, the formation of the lung is dependent on the development of the bronchial vascular system and the pulmonary vascular system that occurs simultaneously.

The steps in lung development are dependent upon inductive cues and crucial interactions between the pulmonary epithelium and the surrounding mesenchyme. The absence of these reciprocal interactions can lead to anatomical anomalies. The regulatory mechanisms and many aspects of the activity of lung mesenchymes are still not well known (Morrisey *et al.*, 2013).

The human lungs arise from the laryngotracheal groove, which evaginates into the surrounding splanchnic mesenchyme. They originate from the anterior surface of the primitive foregut at 5 weeks of gestation in humans and continue to grow for several months after birth (recently reviewed by Warburton *et al.*, 2010; Rankin and Zorn, 2014; Schittny, 2017; Chen and Zosky, 2017).

During the fourth week of embryogenesis, the larynx, trachea, bronchi and lungs begin to form from the respiratory bud, which starts to appear ventral to the caudal portion of the foregut. By the end of the fourth week, a pair of primary bronchial buds evaginate from the laryngotracheal groove (Figure 1). During the fifth week, the left bud gives two secondary bronchial buds while the right bud

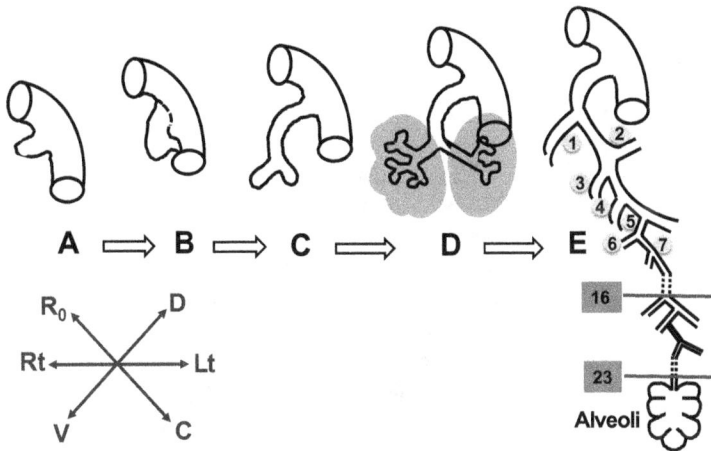

Figure 1. Lung pattern formation and development. (A) The primitive lung anlage emerges as the laryngotracheal groove from the ventral surface of the primitive foregut at 5 weeks' gestation in the human. (B) The primitive trachea separates dorsoventrally from the primitive esophagus as the two primary bronchial branches arise from the lateral aspects of the laryngotracheal groove at 5 or 6 weeks' gestation in the human. (C) The embryonic larynx and trachea with the two primary bronchial branches are separated dorsoventrally from the embryonic esophagus at 6 weeks in the human. (D) The primitive lobar bronchi branch from the primary bronchi at 7 weeks in the human. (E) A schematic rendering of the airway at term in the human. The stereotypical first 16 airway generations are complete by 16 weeks in humans; between 16 and 24 weeks, further branching is nonstereotyped. Alveolarization begins about 20 weeks in humans and is complete by 7 years of age at the earliest. (After West, Burri, Warburton, and others) (adapted from Warburton *et al.*, 2010).

gives three secondary bronchial buds. These secondary bronchial buds give rise to the lung lobes. Over the following week, the secondary buds branch into tertiary buds, which number ten on each side. From the sixth week to the sixteenth week, the major elements forming the lungs appear, except the alveoli. From week 16 to week 26, the bronchi increase in size and lung tissue vascularises highly. Bronchioles and alveolar ducts also develop. Alveolarization begins at around 20 weeks in humans, and continues postnatally till 7 years of age (Figures 1 and 2; Warburton *et al.*, 2010; Schittny, 2017).

In the human lung, 23 generations of airway branching are found (Figure 1). The first sixteen generations branch stereotypically. This stage of branching is completed by 16 weeks, while the remaining

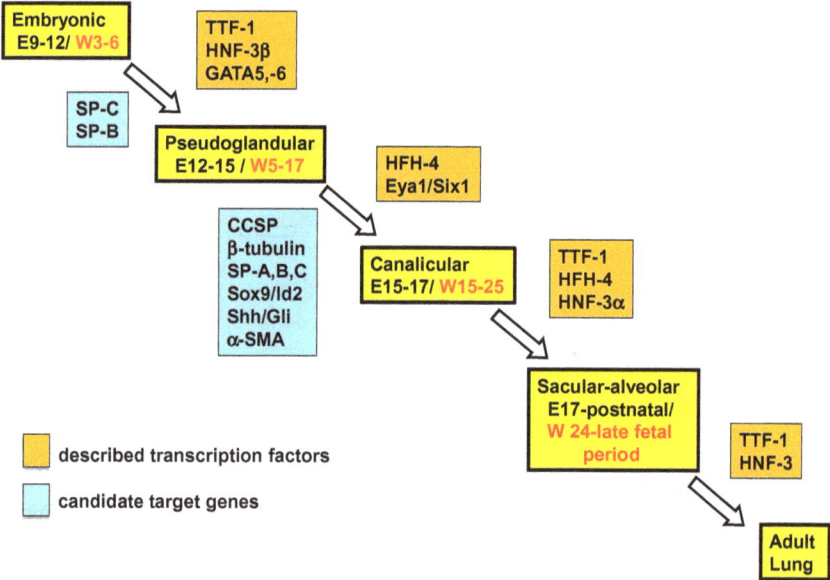

Figure 2.   Timeline of the phases of lung development and some key transcription factors with their specific targets affecting each phase. Timeline for lung development in the mouse is shown in black text in days [E], while for human it is shown in red text in weeks [W]. Despite identical phases of lung development, the timing for each phase is markedly different between mice and human. Transcription factors are shown in an orange box, and their known targets, at the same phase of development, are shown in a blue box.

seven generations are completed almost at 20–24 weeks. Differentiated type I alveolar epithelial cells, which are involved in the process of alveolar blood gas exchange, together with the type II alveolar epithelial cells that produce pulmonary surfactants are formed during lung development. The surfactant reduces the surface tension at the air–alveolar surface, and therefore helps the terminal saccules to expand (Warburton *et al.*, 2010; Schittny, 2017).

1.2 Brief Overview of the Human Respiratory System

The respiratory system consists of two divisions: the upper respiratory tract, which consists of nose and pharynx, and the lower respiratory tract, which consists of larynx, trachea, right and left bronchi, and lungs.

The bronchial tree consists of different parts, such as the primary (main) bronchus, secondary (lobar) bronchus for each lobe, tertiary (segmental) bronchus for each segment, conducting (lobular) bronchiole (1 mm; no cartilage), terminal bronchiole, respiratory bronchiole and alveolar duct as well as the alveoli.

Air first enters our body through the nose or mouth, and then passes through the nasopharynx into the oral pharynx. Then, it passes through the glottis and the trachea into the right and left bronchi, which branch into bronchioles, each of which terminates in a cluster of alveoli. There are 300 million alveoli, which provide a surface area of 160 m² for gas exchange to take place, in a pair of adult lungs.

The pharynx is divided into three parts: nasopharynx, oropharynx and laryngopharynx. The nasopharynx lies posterior to the nasal cavity above the soft palate. The pharyngeal tonsil is in the roof of the nasopharynx. On the lateral wall of the nasopharynx is the opening of the auditory tube. This part of pharynx has a respiratory function as it is a passage for air. The oropharynx is the oral part of the pharynx and lies posterior to oral cavity, extending from below soft palate to the level of vertebra C3. The floor of the oropharynx is formed by the posterior one-third of the tongue. On the lateral wall of the oropharynx are the palatine tonsils. The oropharynx is the part of the pharynx that is a common pathway for air and food. The laryngopharynx is the part of the pharynx that extends from the oropharynx above and continues as oesophagus. It lies behind the opening into the larynx and extends to the level of C6. This part of the pharynx is also a common pathway for air and food.

The human trachea is normally 13 cm long and 2.5 cm in diameter. It has a fibroelastic wall made of U-shaped bars of hyaline cartilage, which keep the lumen patent. The posterior free ends of the tracheal cartilages are connected by a smooth muscle, the trachealis muscle. The trachea commences in the neck below the cricoid cartilage (the 6th sixth cervical vertebra) and ends at the sternal angle by dividing into the right and left bronchi. The bifurcation is called the carina. During deep inspiration, the carina descends to the level of the sixth thoracic vertebra. An enlargement of the thyroid gland in the neck can cause gross displacement or compression of the trachea, while a dilatation of the aortic arch (aneurysm) can also compress the trachea.

Bronchial asthma is a common condition that is characterized by a clear narrowing of the airways. Asthma is caused by a remarkable smooth muscle cell contraction, and an edema of the mucosa as well as accumulated mucus in the lumen of both the bronchi and bronchioles. These pathological changes occur as the result of the rapid histamine release that markedly affects both the caliber and tone of blood vessels (Holgate *et al.*, 2015). In addition, the difficulty of expiration in asthma is most likely because of the bronchioles, which are normally opened during inhalation of air and also remain open during exhalation of air to permit a rapid air outflow (Holgate *et al.*, 2015).

Lungs have a half-cone shape, with a base, apex, two surfaces (costal and mediastinal) and three borders (anterior, inferior and posterior). The lung base sits on the diaphragm, while the lung apex projects above the first rib and into the root of the neck. The hilum is where structures enter and leave the lung. The heart and major vessels indent the mediastinal surface of the lung.

The lungs are surrounded by pleura. Parietal pleura covers the diaphragm, mediastinum and the thoracic wall, while visceral pleura covers the outer surfaces of the lungs and dips into its fissures. The pleural cavity localizes between these two layers and is filled with pleural fluid secreted by the pleural membranes. The pleural cavity functions to provide lubrication, thereby preventing friction during breathing. Parietal pleura is reflected from the mediastinum to the lung to form the visceral pleura. Such an arrangement creates recesses in the pleural cavity between the layers of the pleural reflections.

The parietal pleura consists of the cervical pleura, the costal pleura, the diaphragmatic pleura and the mediastinal pleura. The cervical pleura extends up into the neck, lining the undersurface of the supra-pleural membrane. It reaches a level 1–1.5 inch (2.5–4 cm) above the medial third of the clavicle. The costal pleura lines the inner surfaces of the ribs, the costal cartilages, the intercostal spaces, the sides of the vertebral bodies, and the back of the sternum. The diaphragmatic pleura covers the thoracic surface of the diaphragm. The mediastinal pleura forms the lateral boundary of middle mediastinum.

1.3 Phases of Lung Development

Lung development has been divided into five phases, the embryonic, pseudoglandular, canalicular, terminal saccular and alveolar (Warburton *et al.*, 2010; Schittny, 2017; Chen and Zosky, 2017, Figures 2 and 3 for some histological stages of murine lung development).

During early embryonic development, many signals from the developing mesenchymal tissues act to pattern the development of foregut endoderm in an *anterior–posterior manner*. This *patterning* occurs first into broad domains and then into more specific and restricted regions where the anlages of the prospective organ will bud into the neighboring mesenchymal tissue (Warburton *et al.*, 2010). In addition, *morphological movements* of the developing embryo during embryogenesis can expose the *growing endoderm* to changing patterns of mesenchymal cell-specific gene expression. Consequently, specific regions of the *developing foregut* can

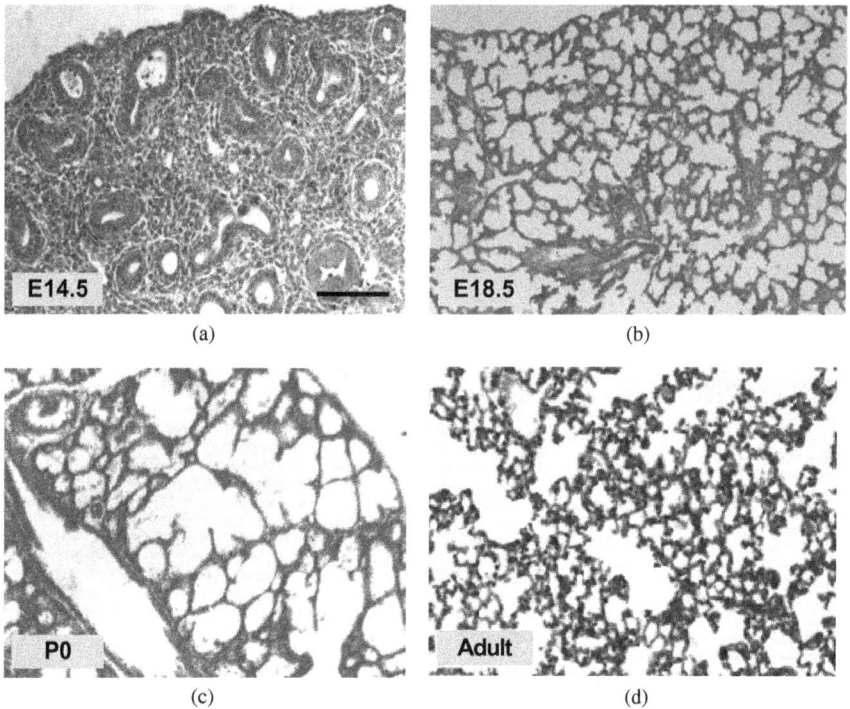

Figure 3.   Hematoxylin and Eosin staining shows histology of murine lung at different developmental stages. (a–d) Embryonic mouse lung develops from pseudoglandular stage (a; E14.5) to canalicular stage, and further terminal sac stage (b–c; e18.5 and P0). Neonatal mouse lungs undergo alveolarization, resulting in the formation of many septa. A mature honeycomb-like structure with alveoli surrounding alveolar ducts conferring normal respiratory structure and function is finally formed, as observed in the adult lung (d). Scale bar: 100 μm (adapted from Warburton *et al.*, 2010).

experience gradients of different morphogens at specific times of foregut development. These characteristically well-organized signaling environments can function to tightly regulate the normal patterning of the developing foregut endoderm. The timing, strength and extent of exposure of these developmental signals can indeed have a significant impact on the differentiation and specification of organ-specific foregut endodermal progenitor cells (Warburton *et al.*, 2010). Identification and characterization of signals and pathways that can promote *lung endodermal development* are, therefore, crucial to determine the generation of many specific types of epithelial cells from undifferentiated pluripotent stem cells, which can provide both basic scientific knowledge and tools for the treatments of a wide range of lung diseases.

1.3.1 *The embryonic phase (3–6 weeks of human pregnancy)*

It starts by the formation of a groove in the ventral lower pharynx, the sulcus laryngotrachealis (28 days of human pregnancy). From the lower part, a bud forms and is called the true lung primordium; this happens after a couple of days (30 days of human pregnancy). Another subdivision occurs leading to the formation of two main bronchi (33 days of human pregnancy). The right lung bud is directed in a caudal direction, while the left smaller bud is directed more laterally. As a result, the asymmetry of the two bronchi, as they present in adults, is already established. On the left, two further buds form while on the right, three further buds appear due to the subsequent divisions of the endodermal branches. These buds correspond to the pulmonary lobes seen later. At the end of the embryonic period, segments of the pulmonary lobes arise.

1.3.2 *The pseudoglandular phase (5–17 weeks of human pregnancy)*

Lungs resemble the development of a tubulo-acinous gland at this developmental stage. The epithelium, which resembles an endocrine gland, undergoes branching morphogenesis. The air-conducting bronchial tree till the terminal bronchioli is set down in this phase (16 generations). Morphometric studies by Kitaoka and co-workers (Kitaoka *et al.*, 1996) have shown that 20 generations are partially present in the lungs by the end of this phase, which means that at this time the respiratory ducts have already been formed.

The cubic epithelial cells are the precursors of the ciliated lung epithelium, and initially function as a coat of the air-conducting bronchial tree. In the 13th week of human pregnancy, the first ciliated epithelial cells can be found (Adamson, 1997). In the respiratory part, the first typically lung-specific cells, type II pneumocytes (alveolar cells) also appear (Kitaoka *et al.*, 1996). Up to the time of birth, the well-developed bronchopulmonary epithelial cells produce the amniotic fluid that is also found in the lungs.

The differentiation of the lungs takes place in a centrifugal direction. Lung epithelial cells differentiate into both cilia-carrying cells and goblet cells in the central air-conducting portions of the lungs. Bronchial glands and both cartilage and smooth muscle cells are localized in the bronchial walls after the 10th week of human pregnancy. The peripheral sections are partially capable of retaining the cubic epithelium that is still little differentiated until far beyond the pseudoglandular phase. This is of importance for the bronchial tree to further proliferate into the surrounding mesenchyme (Warburton *et al.*, 2010; Schittny, 2017).

1.3.3 *The canalicular stage (15–25 weeks of human pregnancy)*

In this phase, the canaliculi branch out of the terminal bronchioli. The terminal bronchioles divide into respiratory bronchioles and alveolar ducts, and the epithelial cells of the airway differentiate into both proximal cuboidal cells and peripheral squamous cells. The respiratory tree expands more in length and diameter, accompanied by further angiogenesis and vascularization. In addition, air spaces that are derived from terminal bronchioles form an acinus.

Remarkable changes in both the epithelium and the surrounding mesenchymal cells are the major features of this canalicular phase. Along the acinus, which develops from the terminal bronchioles, an invasion of capillaries into the mesenchyma occurs. Capillaries that surround the acini form the foundation for the later exchange of gases. The lumen of the tubules becomes wider, and a part of the epithelial cells become flatter (Warburton *et al.*, 2010; Schittny, 2017).

The flattened type I pneumocytes develop from the cubic type II pneumocytes. A sufficient differentiation of the type II pneumocytes into the type I pneumocytes and the proliferation of the capillaries into the mesenchyma mark an important step towards the fetus being able to survive outside the uterus after roughly the 24th week of pregnancy (Warburton *et al.*, 2010; Schittny, 2017).

1.3.4 *The saccular phase (24 weeks to late fetal period in human)*

During this phase, the lung epithelium produces most of the amniotic fluid. In addition, the activity of the type II pneumocytes, which start to secrete the surfactant, can be used for measuring the maturity of the lungs from this phase. The lecithin–sphingomyelin ratio in the amniotic fluid is also determined and increases with fetal age. From the last trimester, clusters of sacs form on the terminal bronchioli, which represent the last subdivision of the passages that supply air (Warburton *et al.*, 2010; Schittny, 2017).

In the bronchial tree, the last generation of air spaces appear at this phase At the end of each respiratory tract passage, smooth-walled sacculi form. It is lined with type I and type II pneumocytes. The primary septa between the sacculi are still thick and contain two networks of capillaries that come from the neighboring sacculi. There is substantial thinning of the interstitium at this stage. This results from apoptosis as well as ongoing differentiation of mesenchymal cells (Hashimoto *et al.*, 2002; Lu *et al.*, 2002). There is a small proportion of both collagen and elastic fibers at this phase, while the interstitial space is rich with cells. This matrix, plays an important role in the differentiation and growth of the epithelium that lies above it (Rannels and Rannels, 1989). At the end of this phase, the interstitial fibroblasts begin the production of extracellular material in the interductal and intersaccular space.

1.3.5 *The alveolar phase (late fetal period to childhood in human)*

The alveolar phase normally begins in the last few weeks of the pregnancy. During this phase, the first alveoli are formed from new sacculi. The formation and development of alveoli continue, and therefore almost 1/3 of the roughly 300 million alveoli are completely developed at birth. The parenchyma that lies between the alveoli forms the primary septa between the alveolar saccule and is composed of a double layer of capillaries (Warburton *et al.*, 2010; Schittny, 2017).

Remarkably, almost all generations of both the respiratory and conducting branches are fully generated at birth. In addition, at birth, the sacculi are thin, smooth-walled sacks and correspond to the later alveolar sacculi. At this stage, the alveolar epithelial cells are more differentiated into mature squamous type I pneumocytes and secretory rounded type II pneumocytes. The capillaries at this stage also grow rapidly in the mesenchyme surrounding the saccules to form a complex double network. Moreover, the lymphatic network is well developed in the lung during this stage. In addition, the alveolarization is affected by many exogenous factors including stretch in fetal airway, oxygen concentration, dexamethasone and retinoic acid. Most of the gas exchange surface is formed during this last stage of lung development (Warburton *et al.*, 2010; Schittny, 2017).

During the alveolar phase, the formation of new septa within terminal sacs is the main step for differentiation of the terminal saccule into alveoli. This includes complex interactions among myofibroblasts in the mesenchyme, adjacent airway epithelial cells and vascular endothelial cells. Multiple developmental processes are important for new septa formation, including cell migration of the myofibroblast progenitors within terminal sac walls as well as controlled proliferation, division, growth and differentiation. Myofibroblasts are smooth muscle precursor cells. These cells can migrate within the developing alveolar septa. Then, they can deposit elastin at the septal tips, which is the early step of development of new secondary septa (Bostrom *et al.*, 1996; Lindahl *et al.*, 1997).

Several molecules and signals regulate the alveolarization. Pdgf-a and its receptor PDGFR-α have a role in new septa formation. Pdgf-a–/– or Pdgf-α–/– mutant mice have a phenotype comprising loss of alveolar myofibroblasts and elastin, failure of alveolar septation, and subsequent development of emphysema due to alveolar hypoplasia (Bostrom *et al.*, 1996; Bostrom *et al.*, 2002). Moreover, in mesenchymal cells undergoing normal and abnormal proliferation, the nuclear factor Phf14 has been found to act as a negative regulator of PDGFRα receptor expression, and thus regulates mesenchymal cell growth (Kitagawa *et al.*, 2012). It has been also reported that PDGF promotes the expression of interleukin-13 (IL-13) in smooth muscle cells of the lung through an oxidant signaling mechanism to regulate smooth muscle development (Bansal *et al.*, 2012). In addition to PDGF-A, there are some other key proteins that mediate cell-to-cell interaction within terminal sac walls. Roundabout (ROBOA receptor) and its ligand SLIT have a role in the control of the migration of non-neuronal cells (Wu *et al.*, 2001). In mice, 24 h before birth, saccular mesenchyme surrounding the airways expresses Slit-2. At the same time, Robo is expressed on the apical aspects of the airway epithelium adjacent to the ligand Slit-2. That suggests interactive roles in pulmonary development (Anselmo *et al.*, 2003). A Robo knockout mice show loss of septation and thickened mesenchyme (Xian *et al.*, 2001).

TGF-β-Smad3 signaling activities in peripheral lung epithelial cells are important for the formation of alveolar septa (Chen *et al.*, 2005; Morikawa *et al.*, 2016). Moreover, abrogation of TGF-β type II receptor specifically in lung epithelial cells leads to reduction of type I alveolar epithelial cells (AECI) and alveolar formation (Chen *et al.*, 2007; Morikawa *et al.*, 2016). In addition, murine mutants lacking fibroblast growth factor receptors 3 and 4 (FGFR3 and FGFR4) fail to undergo normal alveolarization. Also, inhibition of NF-κB transcription factor disrupts both angiogenesis and alveolarization (Iosef *et al.*, 2012).

The distal lung saccules are primitive units of alveoli that are required for respiration at postnatal developmental stages, and rely on alveolar type I cell differentiation (Warburton *et al.*, 2010). A recent study has focused on how distal lung epithelial progenitors develop and form the alveolar type I cells, and how the developmental niche of these cells is instructed by the surrounding mesenchyme (Wang *et al.*, 2016). This study showed that histone deacetylase 3 plays a critical role in these processes and in lung development since its knockout mice have pulmonary hypoplasia with a defective differentiation of alveolar type 1 cells and an impaired lung mesenchymal cell proliferation (Wang *et al.*, 2016). Interestingly, this lung phenotype was correlated with decreased activities of Wnt/β-catenin signaling in lung epithelial cells. Indeed, the inhibition of Wnt signaling results in a defective alveolar type 1 cell differentiation *ex vivo*, a phenotype that can be rescued, at least partially, by the genetic activation of Wnt signaling (Wang *et al.*, 2016). The histone deacetylase 3-regulated Wnt signaling activities are, therefore, important for both alveolar type 1 cell differentiation and lung sacculation processes (Wang *et al.*, 2016).

In summary, the lung has a complex structure of inner surfaces and conducting airways that facilitate the vital process of gas exchange. The process of branching morphogenesis occurs at different developmental stages such as the pseudoglandular stage, the canalicular stage, the saccular stage and the alveolarization stage and leads to the generation of most (90%) of the surface area for gas exchange. Pulmonary alveolarization is critical for any kind of lung regeneration, and it is a lifelong process that continues during childhood and adolescence in mammals (Warburton *et al.*, 2010; Narayanan *et al.*, 2012; Schittny, 2017).

1.4 Genetic Control of the Formation Pattern of Early Lung Anlagen

Several gene products function cooperatively to regulate and define the location of the newly formed laryngo-tracheal groove and control early lung induction. In addition, these genes specifically enable and control dorsal–ventral proximal–distal and left–right axes of the lung during development (Figure 2 and Table 1).

The foregut endoderm can generate several important organs such as the lung, thyroid, stomach, liver and pancreas in the adult mammals. Many studies have uncovered the genetic and molecular programs that regulate development of these endoderm-derived organs, and many growth and transcription regulatory factors that

Table 1. Summary of the major genes and signaling molecules that regulate lung development and patterning, with both their expression pattern and lung/tracheal phenotype in mice (not a comprehensive review).

Gene symbol	Gene name	Expression pattern	Lung/tracheal phenotype in mutant mouse	Reference
Signaling molecule				
Egfr	Epidermal growth factor receptor	Epithelium and mesenchyme	Impaired branching and deficient alveolization	Miettinen *et al.* (1995, 1997), Miettinen (1997)
Fgf18	Fibroblast growth factor 18	Mesenchyme	Deficient alveolization	Usui *et al.* (2004)
Fgf9	Fibroblast growth factor 9	Epithelium and pleura	Impaired branching, reduced mesenchyme	Colvin *et al.* (2001)
			Mesothelial FGF9 is principally responsible for maintaining mesenchymal FGF-Wnt/β-catenin signaling, whereas epithelial FGF9 primarily affects epithelial branching	Yin *et al.* (2011)
Greml	Gremlin 1	Epithelium and mesenchyme	Deficient alveolization	Michos *et al.* (2004)
Hip1	Huntingtin-interacting protein 1	Mesenchyme	Impaired branching	Chuang *et al.* (2003)

(Continued)

Table 1. (*Continued*)

Gene symbol	Gene name	Expression pattern	Lung/tracheal phenotype in mutant mouse	Reference
Shh	Sonic hedgehog	Epithelium	Impaired branching, tracheoesophageal fistula	Litingtung *et al.* (1998)
Tgfb3	Transforming growth factor, β3	Epithelium and pleura	Impaired branching	Kaartinen *et al.* (1995)
Wnt7b	Wingless-related MMTV integration site 7B	Epithelium	Vascular defect, reduced mesenchyme	Shu *et al.* (2002)
Catnnb1	β-Catenin	Epithelium	Impaired branching, proximal/distal specification	Mucenski *et al.* (2003)
Ltbp4	Latent transforming growth factor β binding protein 4	Not reported	Pulmonary emphysema	Sterner-Kock *et al.* (2002)
Wnt5a	Wingless-related MMTV integration site 5A	Mesenchyme and epithelium	Increased branching, tracheal defect	Li *et al.* (2002)
Fgf10	Fibroblast growth factor 10	Mesenchyme	Lung agenesis	Sekine *et al.* (1999)
Fgfr2b	Fibroblast growth factor receptor 2b	Epithelium	Lung agenesis	De Moerlooze *et al.* (2000)
Fgf8	Fibroblast growth factor 8	Not reported	Right pulmonary isomerism	Fischer *et al.* (2002)

TAB1	TGF-B activated kinase-1 binding protein-1	Epithelium	Lung dysmorphogenesis	Komatsu et al. (2002)
Acvr2b	Activin receptor IIB	Not reported	Right pulmonary isomerism	Oh and Li (1997)
Nodal	Nodal	Not reported	Right pulmonary isomerism	Lowe et al. (2001)
Lefty1	Left right determination factor 1	Not reported	Left pulmonary isomerism	Meno et al. (1998)
Traf4	Tnf receptor associated factor 4	Not reported	Tracheal defect	Shiels et al. (2000)
Fgfr3/ Fgfr4	Fibroblast growth factor receptor 3/4	Epithelium and mesenchyme	Defective elastin production, alveolarization defect	Weinstein et al. (1998)
Nog	Noggin	Mesenchyme	Lobation defect	Weaver et al. (2003)
Pitx-2	Paired-like homeodomain transcription factor 2	Not reported	Bilateral isomerism	Kitamura et al. (1999)
Dermo1	Twist homolog 2	Mesenchyme	Impaired branching	De Langhe et al. (2008)
BMP4	Bone morphogenic protein 4	Epithelium and mesenchyme	Abnormal lung morphogenesis with cystic terminal sacs	Bellusci et al. (1996)
Igf1r	Insulin-like growth factor 1 receptor	Not reported	Impaired development	Liu et al. (1993)

(*Continued*)

Table 1. (*Continued*)

Gene symbol	Gene name	Expression pattern	Lung/tracheal phenotype in mutant mouse	Reference
Notch2/3	Notch gene homolog 2/3	Epithelium	Defective myofibroblast differentiation, alveolarization defect	Xu et al. (2009)
PDGFa	Platelet derived growth factor a	Epithelium	Defective myofibroblast elastin production, alveolarization defect	Bostrom et al. (2002)
Timp3	Tissue inhibitor of metalloproteinase 3	Mesenchyme	Reduced number of bronchioles and attenuated alveogenesis	Gill et al. (2003)
Transcription factors				
Nfib	Nuclear factor I B	Epithelium and Mesenchye	Severe lung maturation defects	Hsu et al. (2011)
HuR	RNA-binding protein (RBP)	Epithelium and Mesenchyme	Disrupted morphogenesis of distal bronchial branches	Sgantzis et al. (2011)
Elf3	E74-liketranscription factor-3 (Elf3	Epithelium		Oliver et al. (2011)
NF-κB	NF-κB	Epithlium and Endothlium	Disrupted angiogenesis and alveolarization	Iosef et al. (2012)
Phfl4	Phfl4 transcription factor		Interstitial hyperplasia	Kitagaw et al. (2012)
Foxn4	Foxn4 transcription factor	Mesenchyme	Alveolar defects and reduced septa formation	Li and Xiang (2011)
Cebpa	CCAAT/enhancer binding protein (C/EBP), α	Epithelium	Hyperproliferation of type II cells	Sugahara et al. (2001)
Eya1	Eyes absent	Epithelium and Mesenchyme	Hypoplastic lungs, impaired branching, reduced smooth muscle	El-Hashash et al. (2011b)

Foxa1/Foxa2	Forkhead box A1/A2	Epithelium	Impaired branching, reduced smooth muscle	Wan et al. (2005)
Six1	Sine Oculus	Epithelium and Mesenchyme	Hypoplastic lungs, impaired branching, reduced smooth muscle	El-Hashash et al. (2011c)
Foxf1a	Forkhead box F1a	Mesenchyme	Impaired branching, lobation defect	Lim et al. (2002)
FoxM1	FoxM1 transcription factor	Epithelium and Mesenchyme	Increased FoxM1 expression in respiratory epithelium inhibits lung sacculation and causes Clara cell hyperplasia	Wang et al. (2010)
Hoxa5	Homeobox A5	Mesenchyme	Impaired branching, tracheal defect	Aubin et al. (1997)
Klf2	Kruppel-like factor 2 (lung)	Not reported	Impaired sacculation	Wani et al. (1999)
Mycn	Neuroblastoma myc-related oncogene 1	Epithelium	Impaired branching	Moens et al. (1992)
Trp63	Transformation-related protein 63	Epithelium	Tracheobronchial defect	Daniely et al. (2004)
Titf1	Thyroid transcription factor 1	Epithelium	Loss of distal lung fate, impaired branching, tracheoesophageal fistula	Kimura et al. (1996)
Nfib	Nuclear factor I/B	Epithelium and mesenchyme	Sacculation defect	Steele-Perkins et al. (2005)

(Continued)

Table 1.   (*Continued*)

Gene symbol	Gene name	Expression pattern	Lung/tracheal phenotype in mutant mouse	Reference
Sox11	SRY-box-containing gene 11	Epithelium	Hypoplastic lung	Sock *et al.* (2004)
Tcf21	Transcription factor 21 (Pod1)	Mesenchyme	Impaired branching	Quaggin *et al.* (1999)
Rarb/Rara	Retinoic acid receptor α/β	Epithelium and mesenchyme	Left lung agenesis and right lung hypoplasia	Mendelsohn *et al.* (1994)
Pitx2	Paired-like homeodomain transcription factor 2	Mesenchyme	Right pulmonary isomerism	Lin *et al.* (1999)
Foxj1	Forkhead box J1	Epithelium	Left–right asymmetry, loss of ciliated cells	Brody *et al.* (2000)
Gata6	GATA-binding protein 6	Epithelium	Impaired sacculation	Yang *et al.* (2002)
Gli2/Gli3	GLI-Kruppel family member GLI2/GLI3	Mesenchyme	Lung agenesis	Motoyama *et al.* (1998)
Ascl1	Achaete-scute complex homolog-like 1	Neuroendocrine cells	Loss of neuroendocrine cells	Ito *et al.* (2000)
Erm	Ets variant gene 5	Epithelium	Impaired type I cell formation	Liu and Hogan (2002), Liu *et al.* (2003)

Gene	Description	Tissue	Phenotype	Reference
Wnt2/2b	Wingless-related MMTV integration site 2/2b	Mesenchyme	Complete lung agenesis	Goss et al. (2009), Harris-Johnson et al. (2009)
Alk3	Aurora-like kinase	Epithelium	Retardation of lung branching, reduced cell proliferation and differentiation	Sun et al. (2008)
Others				
Fstl1	a BMP 4 signaling antagonist,	Mesenchyme	Defective alveolar maturation and malformed tracheal rings and reduced ring number	Geng et al. (2011).
Eln	Elastin	Mesenchyme	Deficient alveolization	Wendel et al. (2000)
Cby	Wnt/β-catenin antagonist Chibby (Cby)		Altered lung morphogenesis, epithelial cell differentiation and mechanics	Love et al. (2010)
Lmnb1	Lamin B1	Epithelium and mesenchyme	Deficient alveolization	Vergnes et al. (2004)
Lama5	Laminin α5	Epithelium and pleura	Defective lobation	Nguyen et al. (2002)
Pcaf	p300/CBP-associated factor	Epithelium and mesenchyme	Defective proximal and distal epithelial cell differentiation	Shikama et al. (2003)
Adam17	A disintegrin and metallopeptidase domain 17	Epithelium	Impaired epithelial differentiation, impaired branching	Zhao et al. (2001), Peschon et al. (1998)
Crh	Corticotropin-releasing hormone	Epithelium	Defective epithelial and mesenchymal maturation	Muglia et al. (1999)

(Continued)

Table 1. (*Continued*)

Gene symbol	Gene name	Expression pattern	Lung/tracheal phenotype in mutant mouse	Reference
Pthlh	Parathyroid hormone-like peptide	Epithelium	Deficient alveolization	Rubin *et al.* (2004)
Itga3	Integrin α3	Epithelium	Impaired branching	Kreidberg *et al.* (1996)
Cutl1	The transcriptional repressor CDP Cut-like 1 (Cutl1)	Epithelium	Impaired epithelial differentiation	Ellis *et al.* (2001)
RXRa	Retinoic X receptor alpha	Epithelium and mesenchyme	Decrease in alveolar surface area and alveolar number	McGowan *et al.* (2000)
Tmem16a	Transmembrane protein 16a	Epithelium	Abnormal tracheal cartilages resulting in tracheomegaly	Rock and Harfe (2008) Rock *et al.* (2008)
TACE	Tumor necrosis factor-α converting enzyme	Not reported	Failure to form saccular structures	Zhao *et al.* (2001)
PDGFa	Platelet-derived growth factor a	Epithelium	Defective myofibroblast elastin production, alveolarization defect	Bostrom *et al.* (2002)
Na/K ATPase	Sodium/Potassium ATPase	Epithelium	Failure to absorb fetal lung liquid, causes significant respiratory distress and neonatal lethality	Hummler *et al.* (1996)
Lfng	Lunatic Fringe	Epithelium	Impaired myofibroblast differentiation and alveogenesis	Xu *et al.* (2010)

are critical for specification, commitment and differentiation of these organs were determined (Warburton *et al.*, 2010; Rankin and Zorn, 2014; Kim and Shivdasani, 2016; Zaret, 2016; Villasenor and Stainier, 2017; McCracken and Wells, 2017). Despite this successful progress, essential questions remain unanswered such as what are the transcriptional repressive molecular mechanisms that are essential for maintaining and promoting the identity and/or differentiation of the organ-specific endoderm?

One of the key transcription families in endoderm specification and differentiation is the forkhead box transcription factor (Fox) family, which is a group of transcriptional regulatory factors. These factors can be defined through their characteristic and common forkhead domain for DNA binding. Fox family members can be divided into small subfamilies on the basis of their similarities to other protein domains (outside the forkhead domain).

The members of Foxp subfamily are effective inhibitors of gene expression through their interaction with certain complexes that act to repress chromatin remodeling, including NuRD (Chokas *et al.*, 2010). Foxp transcription factors have the ability of restricting the programs of cell lineage differentiation that remarkably depends on their potent capacities to inhibit the expression of certain lineage-specific genes (Chokas *et al.*, 2010). The Foxp subfamily is made of 4 members, Foxp1/2/3/4. These Foxp1/2/3/4 members have several conserved domains that are required for repressing gene transcription (Li *et al.*, 2004). In addition, the expression of Foxp1/2/4 is detected in several endoderm cell lineages such as the lung, pancreas and thyroid (Lu *et al.*, 2002 and Spaeth *et al.*, 2015), while Foxp3 is expressed in the T cell lineage in the hematopoietic system (Hori *et al.*, 2003). Furthermore, several studies have shown a cooperative role of Foxp1/4 and Foxp1/2 in regulating lung endoderm differentiation (Li *et al.*, 2012 and Shu *et al.*, 2007), and all three family members are expressed in an overlapping pattern in the lung endoderm during development (Lu *et al.*, 2002). Recently, the combined role for Foxp1/2/4 in pancreatic alpha cell proliferation has been reported (Spaeth *et al.*, 2015).

In these studies, deletion of more than one member of the Foxp family was essential to detect a phenotype during organ formation and development, suggesting clear cooperativity and redundancy between the Foxp family members.

The Foxp1/2/4 transcription factors are required for major processes in lung development such as the proper commitment and/or differentiation of lung-specific endodermal lineages (Lia *et al.*, 2016). Foxp1/2/4 function together to repress some key transcriptional factors that are not normally expressed in the lung-specific endodermal cells such as Pax2/8/9 and Hoxa9-13 (Lia *et al.*, 2016). This study suggests an important role for Foxp1/2/4 transcription factors in the regulation of the expression levels and activities of certain cell lineage-specific genes that are essential for proper tissue/organ development.

Furthermore, during late lung maturation, Foxp1/2 transcription factors are critical for alveolar type 1 epithelial cell development (Shu *et al.*, 2007), while Foxp1/4 are important for both development and regeneration of airway secretory cells after naphthalene-induced lung injury (Li *et al.*, 2012). Indeed, loss of Foxp1/4 transcription factors in mice can lead to the ectopic activation of the goblet cell-specific regulatory genes, as well as the abnormal differentiation into goblet cells in the lung airways that is partially mediated through de-repressing the anterior gradient 2 (Li *et al.*, 2012).

The similarity in Foxp1/2/4 expression pattern and the protein sequence in some tissues such as the lung suggests a certain levels of redundancy or cooperativity in their regulation of gene expression. Some studies on pancreatic alpha cells have shown that the inactivation of Foxp1/2/4 can result in a reduction in the proliferation and mass of these pancreatic cells, which leads to hypoglycemia in adult mice (Spaeth *et al.*, 2015). This suggests that only low expression and activity levels of Foxp are needed for alpha cell proliferation (Spaeth *et al.*, 2015). This contrasts with what happens in the lung where a complete deletion of both Foxp1 and Foxp4 genes leads to significant defects in both the development and regeneration of the airway epithelium (Li *et al.*, 2012).

Several other factors are also critical for the early induction of laryngotracheal groove and both lung symmetry and development. For example, both retinoids and their receptors play important roles in the early lung development since the induction of the laryngotracheal groove is prevented by a retinoic acid deficiency or by compound null mutations of retinoid receptors. Furthermore, the left–right asymmetry of the lung is controlled by of several genes and proteins such nodal, Lefty-1, 2, and Pitx-2. The isomerism of lung is found in Pitx2 null mutant mice, while Lefty-1-/- mutants have bilateral single-lobed lungs (Warburton *et al.*, 2010; Rankin and Zorn, 2014; Schittny, 2017).

Despite the fast-growing research on lung development, little is known about the molecular mechanisms of initial specification of tissue-specific endoderm progenitors in the anterior foregut. Wnt signaling has been shown to play a critical role in specifying Nkx2-1+ lung endoderm progenitors. It is one of the genetic models that causes complete respiratory agenesis (Goss *et al.*, 2009; Harris-Johnson *et al.*, 2009). Interestingly, Foxp2 expression, which is important for the generation of the thyroid and lung, can demarcate the same Nkx2-1 expression-marked region in the anterior foregut endoderm (Sherwood *et al.*, 2009). In a recent research, Lia *et al.* (2016) have determined the critical importance of cooperative Foxp1/2/4 function in regulating early lung development, and the morphological presence of a respiratory organ suggests that these three transcription factors may not be essential for specification of lung endoderm progenitors. Foxp1/2/4 are, therefore, important regulators of the early development of lung endoderm progenitors and act, at least partially, by inhibiting the non-lung gene expression program(s) (Lia *et al.*, 2016).

During the early gut morphogenesis, the earliest endodermal signal that is essential is the hepatocyte nuclear factor (HNF/Fox) transcription factors and GATA, which are zinc finger proteins recognizing the GATA DNA sequence. In addition, GATA-6 is important for the activation of the lung developmental process in the foregut endoderm, while Foxa2 is required for gut tube closure.

For the endoderm, Hnf-3/Foxa2β is regarded as a survival factor, and its expression is induced by sonic hedgehog (SHH) signaling (Kim and Shivdasani, 2016; Zaret, 2016; McCracken and Wells 2017). Furthermore, Tbx4 transcription factor can induce ectopic bud formation in the esophagus by activating the expression of Fgf10 growth factor (Sakiyama *et al.*, 2003).

Beta-catenin and Wnt2/2b signaling are also necessary to specify lung progenitors in the foregut (Goss *et al.*, 2009; Harris-Johson *et al.*, 2009). Wnt2/2b mutant mouse embryos lack the expression of the earliest lung endoderm genetic marker, Nkx2.1 gene, and exhibit complete lung agenesis. This phenotype is recapitulated by an endoderm-restricted deletion of beta-catenin, while conversely conditional expression of an activated form of beta-catenin results in the expansion of Nkx2.1 expression domain into esophagus and stomach epithelium, suggesting the importance of the canonical Wnt2/2b signaling pathway in the specification of lung endoderm progenitors within the foregut (Goss *et al.*, 2009, Harris-Johson *et al.*, 2009).

The mesenchyme provides crucial signals for respiratory lineage specification. The lung mesenchyme is indeed an important source of both inductive signals and cues for the development of the neighboring epithelium. At embryonic day E8.25, the specification of the respiratory system occurs in the ventral anterior foregut endoderm, as indicated by the expression of Nkx2-1 (also named Ttf1) in mice (Minoo *et al.*, 1999; Serls *et al.*, 2005). Several studies on respiratory lineage specification implicate the surrounding ventral mesenchyme as a crucial source of signals, including, WNT, TGF beta, FGF, BMP, and Retenoic Acid (RA), that direct endodermal expression of Nkx2-1 in a temporal and spatial context-dependent fashion (Que *et al.*, 2006; Chen *et al.*, 2007). Indeed, the FGF signaling pathway is critical for early lung formation as well as for early Pax2 expression in the developing thyroid, while RA is required for lung but not for thyroid differentiation in *Xenopus* after gastrulation (Wang *et al.*, 2011).

The early lung development is dependent on several recently identified mediators of mesenchymal–epithelial interactions and specific molecular mechanisms initiated within the developing lung mesenchyme. Combined mesenchymal expression of Wnt2 and Wnt2b as an example of these signals has been shown to be needed for the expression of Nkx2-1 and respiratory lineage specification (Goss *et al.*, 2009). However, in the mesenchyme the upstream factors that control the expression of these signals are less clear. A study on *Xenopus* has shown that morpholino knockdown of Osr1 and Osr2, a pair of transcription factor genes, caused loss of Wnt2b expression in the mesenchyme (Rankin *et al.*, 2012). In addition, genetic inactivation of Tbx5 before respiratory specification in mice led to unilateral loss of Nkx2-1 expression in the prospective pulmonary epithelium and both reduced Wnt2 and led to loss of Wnt2b expression in the mesenchyme (Arora *et al.*, 2012). Collectively, these data suggest that in the lung mesenchyme Osr1/2 and Tbx5 are critical for normal Wnt2 and Wnt2b expression and subsequent specification of the respiratory foregut epithelium. Although it was suggested that specification signals from the mesenchyme can control Nkx2-1 expression via transcription factors (Domyan *et al.*, 2011), another study showed that these specification signals can also function through epigenetic mechanism(s) (Herriges *et al.*, 2014). The later study also showed that the long non-coding RNA, NANCI, is regulated by mesenchymal WNT signaling and expressed in the ventral foregut, in which it acts as a positive regulator of Nkx2-1 expression (Herriges *et al.*, 2014).

1.5 Distal Airway Branching Morphogenesis

The lung airway in vertebrates shows a characteristic tree-like branched structure, which is produced by many repeats of tip splitting that is known as branching morphogenesis. Despite intensive research on branching morphogenesis in different branching organs such as kidney, placenta and lung, the precise mechanisms of pattern formation and modeling of the lung branching morphogenesis are not well understood (Warburton *et al.*, 2010; Miura, 2015).

In the mammalian lung, the bronchial tree is regarded as a paradigm of biological complexity (Goldberger and West, 1987).

Abundant geometric measurements, made from polymeric casts of bronchial trees in several mammalian species, have shown that the airways are formed of many generations of dichotomous branches, arranged in a highly ordered, space-filling tree (West, 1987). In humans, this tree has, on average, 23 generations (Weibel, 1984). That is, a hierarchy of 22 branch points separates each distal alveolus, from the trachea. Interestingly, the overall branching pattern is highly stereotyped, so the geometry is quite similar amongst individuals (Metzger, 2008).

During lung branching morphogenesis, the intra-pulmonary airway branching process that occurs distal to the primary bronchi is apparently complex as it proceeds distally and during the increase of the individual branch number. When the laryngotracheal complex becomes well-formed, the branching of distal airways will be actively driven by a specific master branch-generator routine, to a rotational orientation subroutine that is able to direct the orientation of the branches around the axis of the airway, and finally to a bifurcation subroutine (Metzger *et al.*, 2008; Warburton, 2008; Warburton *et al.*, 2010; Figure 4).

During early murine lung formation, the branching morphogenesis of the bronchi can be anatomically parsed into some simple and easily recognized geometric forms (three in number) that are called domain branching, planar and orthogonal bifurcation (Metzger *et al.*, 2008). Then, these simple geometric forms are normally repeated iteratively during the process of lung formation to produce different characteristic arrangements of complex pulmonary branches. These arrangements are known as bottlebrush array and planar array as well as rosette array (Metzger *et al.*, 2008; Warburton, 2008; Warburton *et al.*, 2010; Figures 4 and 5).

The bottlebrush array refers to the sequential proximal to distal emergence of secondary branches along the airway. The bottlebrush array is then reoriented to form a second row of branches at right angles to the first row. The planar array and rosette terms are known to describe the patterns that are formed by the sequential bifurcation of the tip of tertiary, secondary and buds at right angles to each other

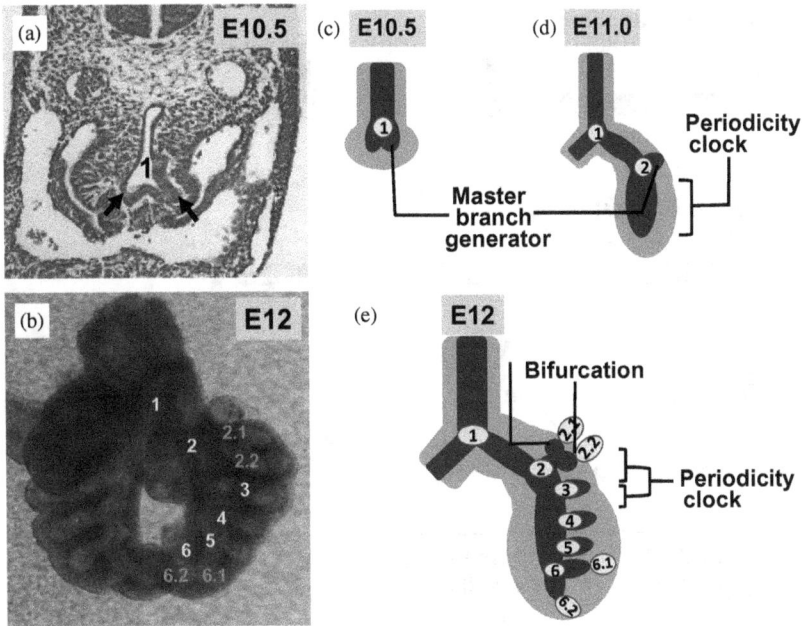

Figure 4.   Formation of the airways in a sequential manner by reiterating a few, relatively simple sets of genetic instructions. In (a–e), which are figures and drawn after Warburton (2008), Warburton *et al.* (2010), and El-Haashash *et al.* (2011b), a master branch generator, a periodicity clock and a bifurcator program are shown as controlling the layout of the mainstem and lobar branches. At embryonic day (e) 10.5, (a,c) the primary bronchial branch (1) forms, followed by (d) the development of the left upper-lobe branch (2) by E11, and then (b,e) the first two segmental branches of the left upper-lobe branch (2.2 and 2.3) form and the subsequent formation of branches 3–6 occurs by E12. The master branch generator is active throughout these events, and the inferred sites of action of the periodicity clock and bifurcator subroutines are shown (adapted from Warburton *et al.*, 2010).

(Warburton *et al.*, 2010). The repetition process of these simple modules of branching, with the precise hierarchical control and coupling of them, may explain the ability of the genome to encode the complex and well-organized stereotypic pattern of formation that characterizes the early bronchial branches in the developing lung using simple and effective genetic modules (Warburton *et al.*, 2010).

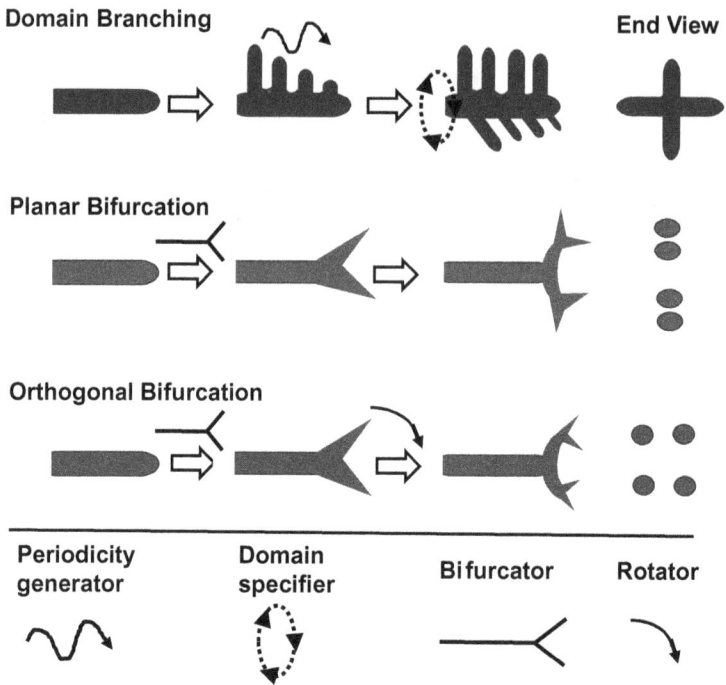

Figure 5.   Formation of complex patterns of peripheral branching. Following the views of Metzger *et al.* (2008), a series of inferred genetic subroutines are shown, all driven by one master branch generator, shown as giving rise to domain or "bottle brush" branching along the lateral proximodistal axis of the main stem bronchi, which can then be rotated at right angles to give rise to a second rank of branches. Then, in subsequent rounds of branching, arising from the tips of the primary and secondary branches, it is shown how the same relatively simple periodicity generator, domain specifier, bifurcator and rotator subroutines can give rise to apparently more complex patterns of peripheral branching to achieve an ever larger number of space-filling terminal branches (adapted from Warburton *et al.*, 2010).

1.6 Alveolar Septum Formation

The final stage of lung development is the alveolarization process that begins before birth and continues through the newborn period and childhood, with overall body growth. During the first phase of alveolarization, known as primary septation, the airway saccules become thinner, bringing the airway epithelium in close contact with the lung vasculature, thereby forming a basic unit for gas exchange. Subsequently, the alveoli mature through the process of secondary septation, in which multiple septa grow inward from the walls of each alveolus. Each septum contains two epithelial layers with a central core that contains single capillary and mesenchymal components that include myofibroblasts and elastin as well as other extracellular matrix components. The migration and differentiation of myofibroblasts in the alveolar septa are critical (Branchfield *et al.*, 2016), but this remains a poorly understood aspect of lung development, as does the formation of lipofibroblasts (McGowan and Torday, 1997).

Within the lung, the process of secondary septation greatly increases the surface area for gas exchange. It occurs immediately prenatally and continues postnatally, thus providing an attractive target for designing therapeutic interventions for patients with CDH and other neonatal lung diseases. Deficient alveolarization has been observed in CDH, resulting from nitrofen administration. This can be improved by treatment with corticosteroids during the antenatal period (Suen *et al.*, 1994).

Human studies have proven more challenging, likely because of pragmatic clinical difficulties in obtaining appropriate pathology specimens, because most infants with CDH undergo extensive ventilatory interventions that disrupt the natural architecture of the lung. Major bronchial subdivisions are normal in human autopsy specimens of hypoplastic CDH lungs; however, the number of intermediate bronchial branches is often reduced severely in the ipsilateral lung but to a lesser extent in the contralateral lung, which comparatively maintains the potential of faster growth and complete differentiation during gestation (Coleman *et al.*, 2015; Phithakwatchara *et al.*, 2015). The prevalence of left-sided hernias suggests that left–right asymmetry may also be an important component of the pathophysiology of the hypoplasia.

Two more processes are necessary in septum differentiation to have a septum with final mature function and morphology. One is the maturation of the double capillary network into a single capillary bed, and the other is thinning out of the septal mesenchyme. Thinning of the mesenchymal tissue involves apoptosis within the postnatal lung mesenchyme. In addition, during this rapid alveolarization phase, there is a substantial decrease in the number of interstitial myofibroblasts that results from increased apoptosis (Awonusonu *et al.*, 1999; Schittny *et al.*, 1998). Interestingly, the immature lung contains at least two morphologically distinct fibroblast populations: lipid-filled interstitial fibroblasts (LFIF) and non-LIF (NLFIF).

After alveolarization, apoptosis occurs preferentially in the LFIF, which is correlated with a diminished expression of insulin-like growth factor I receptor (Igf-IR) mRNA and cell surface protein (Srinivasan *et al.*, 2002). This thinning of the previously thickened immature interstitium occurs simultaneously with the ongoing expansion of the epithelial, blood vessel and airspace compartments within the rapidly developing septa. In addition, a mature capillary bed is vital for the proper function of alveoli, although the mechanism for its formation is not completely understood (reviewed in Warburton *et al.*, 2010; Morrisey and Hogan, 2010; Schittny, 2017).

During lung development, vascular endothelial growth factor (VEGF) isoforms and its receptors have been identified as being important for endothelial survival and proliferation in the alveolar wall. Inhibition of VEGF signaling results in abnormal lung vascular growth and reduced alveolarization (Kasahara *et al.*, 2000; Warburton *et al.*, 2010; Morrisey and Hogan, 2010; Schittny, 2017).

Finally, the new septum differentiates into a functional respiratory membrane that consists of type I alveolar epithelial cells, basement membrane and capillary endothelial cells. The respiratory membrane provides a short distance for diffusion (around 1 micron) that helps in facilitating gas exchange. It is estimated that about 50 million alveoli are present in term human neonatal lung. However, by age 7 to 8, when the alveolarization is largely complete, the number of alveolar units in the human lung has increased about six-fold, to reach around 300 million alveoli (Warburton *et al.*, 2010; Morrisey and Hogan, 2010; Schittny, 2017).

1.6.1 *Factors and mechanisms regulating the alveolar septum formation*

Retinoic acid (RA) is produced during vitamin A metabolism, which involves successive oxidative reactions of dietary precursors including carotenoids and retinyl esters (Das *et al.*, 2014). It is an active metabolite of vitamin A. Vitamin A deficiency has long been known to injure lung and has been shown to impair function of rat type pneumocytes (McGowan *et al.*, 2000). These evidences suggest that RA may play an important role in alveolar development. Morphological studies have demonstrated that RA signaling is essential for lung development (Malpel *et al.*, 2000). Retinoids are crucial mediators of prenatal and postnatal lung development (Massaro *et al.*, 2003; Yang *et al.*, 2003). A recent animal study demonstrated that alveolar formation and regeneration can be rescued using postnatal RA treatment in calorie-restricted developing rat lungs (Londhe *et al.*, 2013).

In a human study, in extremely low-birth weight infants, vitamin A supplementation decreased the incidence of bronchopulmonary dysplasia (Tyson *et al.*, 1999). In rats, Retinoid acid (RA) has been shown to increase the number of alveoli (Massaro and Massaro, 1996) and partially rescue a block in alveolar formation induced by dexamethasone (Massaro and Massaro, 2000). RA has also been reported to ameliorate the septal destruction associated with elastase-induced emphysema in adult rats (Massaro and Massaro, 1996). In the RAR-γ gene deletion mouse model, there is a developmental defect in alveolar formation most consistent with a defect in elastin deposition. Also, the additional deletion of one retinoid X receptor (RXRα) allele results in a decrease in alveolar number and alveolar surface area (McGowan *et al.*, 2000). Moreover, retinoids affect multiple cellular functions that are involved in alveolar septal formation such as migration, proliferation and temporal differentiation of cells (Chytil, 1996).

Furthermore, other studies have shown that myofibroblast development is regulated by Wnt3a, which induces myofibroblast differentiation by upregulating TGF-β signaling through SMAD2 in a β-catenin-dependent manner (Carthy *et al.*, 2011). Furthermore, targeted inactivation of Foxn4 causes reduced septa in the distal lung, thinned alveolar walls and dilated alveoli. These alveolar defects may result from reduced surfactant protein B expression and decreased platelet-derived growth factor-A signaling (Li and Xiang, 2011).

1.7　Development of Various Lung-Specific Cell Types

The lung has many specific cell types that compose the proximal airway, distal airway and alveoli. The proximal airway contains pseudostratified epithelium that is made of a basal layer containing both a basal cell layer and a luminal layer that is composed of Clara-like cells (secretory club cells), human goblet cells, ciliated cells and neuroendocrine cells that form aggregates of neuroendocrine bodies (Asselin-Labat and Filby, 2012). The distal airway contains both small bronchi and bronchioles and is composed of a single epithelial layer that consists of neuroendocrine cells, ciliated cells, Clara cells and $p63^+Krt5^+$ basal cells. Clara cells are more abundant in the distal airway compared to ciliated cells. In addition, the distal airway contains more neuroendocrine cells than in the trachea (Asselin-Labat and Filby, 2012). The alveoli exist at the most distal lung region. They consist of the following epithelial cell types: alveolar epithelial type II cells (AEC II) that have secretory vesicles containing surfactant, and alveolar epithelial type I cells (AEC I) that are essential for gas exchange in the lung (Asselin-Labat and Filby, 2012).

During lung development, the early embryonic lung epithelium comprises progenitor cells that co-express many phenotypic markers characteristic of mature cell lineages (Wuenschell *et al*, 1996). The first specific cell lineage to become recognizable by immunohistochemistry is the neuroendocrine lineage, closely followed by the Clara cell lineage. At least 40 specific types of cells are differentiated during embryonic lung development. The epithelial cell lineages become arranged in a distinct proximal–distal spatial pattern in the airways (Warburton *et al.*, 2010; Morrisey and Hogan, 2010; Schittny, 2017).

Cartilage lies outside the submucosa and is reduced in amount as the caliber of the bronchi decreases. Although cartilage is present in the bronchi, it is absent in the bronchioles. Two major cell components of the proximal bronchial epithelium are identified. They are pseudostratified ciliated columnar cells and mucous (goblet) cells. Both cell types are derived from basal cells, but they are ciliated cells predominate in number. Goblet cells release mucus granules into the bronchial lumen, which act to trap particulate matters and effectively prevent drying of the bronchial walls. Mucous cells begin to mature around 13 weeks of gestation in humans, when the mature ciliated columnar cells are already present. The molecular markers for mucous cells are mucins (MUC5B, 5A, 5C). The beating of cilia results in a cephalad movement of the mucus blanket, thereby cleaning and protecting the airway. In the case of cystic fibrosis, cilia movement is disabled due to the thick mucous layer that is caused by mutation of the cystic fibrosis transmembrane conductance regulator (Cftr) gene, which encodes a transmembrane Na+ ion transporter protein. This phenotype also makes the airway surface vulnerable to microbial infection. In chronic airway injury, repair or fully experimental exposure of the epithelium to IL-9 before it becomes fully differentiated both result in goblet cell hyperplasia. Indeed, exposure to IL-9 results in increased lysozyme and mucus production by the epithelia (Vermeer *et al.*, 2003). In addition, IL-4, IL-13 and allergens aid the release of TGF-a, which is a ligand for the epidermal growth factor receptor that also stimulates goblet cell differentiation and fibroblast proliferation (Lordan *et al.*, 2002; Warburton *et al.*, 2010; Morrisey and Hogan, 2010; Schittny, 2017).

Three different types of cells in bronchial submucosal glands have been identified. Myoepithelial cells surround the gland, while mucous cells (pale cytoplasm) and serous cells (basophilic cytoplasm) produce mucins. These different secreted mucins can mix with both IgA and lysozyme on the airway surface. In addition, Kulchitsky cells exist on the airway surface next to bronchial glands. The precise functions of Kulchitsky cells are unclear, but it is believed that they are pulmonary neuroendocrine cells (PNEC) that produce a variety of peptide hormones such as calcitonin and serotonincalcitonin. Their fingerlike cytoplasmic extensions usually reach the airway lumen. Moreover, Kulchitsky cells express the markers gastrin-releasing peptide (GRP), calcitonin gene-related peptide (CGRP) and chromogranin that is related to certain lung neoplasms (i.e. small cell carcinoma and carcinoid tumors). However, PNEC differentiate earlier, by 10 weeks of gestation in humans, and are also the first airway epithelial cells to be fully differentiated in mouse by E16 (Warburton *et al.*, 2010; Morrisey and Hogan, 2010; Schittny, 2017).

Clara cells are found in the distal bronchiolar airway epithelium that normally lacks mucous cells. They produce a mucus-poor, watery proteinaceous secretion. They assist with clearance and detoxification as well as reduction of surface tension in small airways. The most important cellular marker of Clara cell is Clara cell-specific protein (CC10, CCSP or uteroglobin). Moreover, Cytochrome P450 reductase and CC10 can also be used as cellular markers for Clara cells that begin to mature during the 19th week of human pregnancy. Notably, only a small number of mucin-positive cells are present in the lung airway of normal mice. However, numerous mucus cells, which are derived from Clara cells with excessive mucin production or reduced mucin secretion, can be found during mucus metaplasia (Evans *et al.*, 2004), and some of mucus concentration and secreted mucins such as Muc5ac and Muc5b contribute to the pathogenesis of muco-obstructive lung disease (Livraghi-Butrico *et al.*, 2017). Another study has demonstrated that increased expression of FoxM1 transcription factor in respiratory epithelium causes Clara cell hyperplasia and inhibits lung sacculation (Wang *et al.*, 2010).

Most of the alveolar surface is normally covered by type I epithelial cells. These flat cells are believed to be terminal differentiated cells, expressing several specific molecular markers, such as T1a and aquaporin 5. T1a is a differentiation marker gene of lung alveolar epithelial type I cells. It is developmentally regulated and encodes an apical membrane protein of unknown function. Upon deletion of T1a protein, type I cell differentiation is blocked. Homozygous T1a null mice die at birth of respiratory failure, and their lungs cannot be inflated to normal volumes (Ramirez *et al.*, 2003). On the other hand, Aquaporin 5 is a water-channel in type I epithelial cells. A knock-in of Cre-ERT2 into the Aqp5 locus has been recently reported, which is a major advance because these mice will be useful not only to target gene deletion in type I cells but also for lineage tracing of the type I cells under development, injury and repair (Warburton *et al.*, 2010; Morrisey and Hogan, 2010; Schittny, 2017).

Type I epithelial cells only account for 40% of the total number of airway epithelial cells, even though >95% of the alveolar surface area is covered by this type of B cell. The other 60% of alveolar epithelial cells are rounded cells that cover only 3% of the alveolar surface, named type II pneumocytes. Type II pneumocytes are plump or cuboidal and have a finely stippled cytoplasm and surface microvilli. These cell types produce surfactant phospholipids and proteins that reduce the surface tension in the lung. This prevents atelectasis at end-expiration as it equalizes pressure, stabilizes and maintains alveoli in an open position despite the variation in alveolar size. Four surfactant proteins (Sftp), SP-A, B, C and D, also play critical roles in maintaining lung function. SP-A and SP-D contribute to host defense in the lung, whereas SP-B and SP-C contribute to the surface tension-lowering properties of the lipoprotein complex termed pulmonary surfactant (Whitsett and Weaver, 2002). In addition, surfactant protein C (SP-C) is a common cellular marker for type II cells. Furthermore, Type II cells are also capable of regeneration and replacement of type I cells after injury (Warburton *et al.*, 2010; Morrisey and Hogan, 2010; Schittny, 2017).

Another component of the cells in alveoli is alveolar macrophages, which play a crucial role in the host defense mechanism as a major cellular sentinel in the lung alveolar space. They are mainly derived from blood monocytes and are a part of the mononuclear phagocyte system. Alveolar macrophages represent a small percentage of the cells in alveoli, and their turnover rate becomes slow when they get into the lung (Warburton *et al.*, 2010; Morrisey and Hogan, 2010; Schittny, 2017).

1.8 Summary and Conclusions

The respiratory system in mammals arises from the growing anterior foregut through sequential morphogenetic events that are tightly regulated by reciprocal interactions between the developing endoderm and mesoderm. The lung contains two well-organized, highly branched, tree-like systems, which are the airways, and the vasculature that can fulfill the important gas exchange tasks. These well-organized systems develop in a tightly coordinated way from the early developing primary bud stage to the generation of numerous units of alveolar gas exchange.

During early lung development, the formation and expansion of the gas exchange surface area follows the formation of the conducting airways. The process of alveolarization continues after birth until young adulthood. Both the right and left lungs have their own anlage that develops as an outpouching of the foregut during organogenesis. During the pseudoglandular phase, the lung bud undergoes a repetitive and tightly regulated process of both outgrowth and branching morphogenesis that leads to the formation of the future lung airways, while both the pulmonary epithelial cell differentiation is visible and the bronchioalveolar duct junction exist during the canalicular phase and their location of is constant throughout life. Remarkably, the first gas exchange may occur towards the end of the canalicular phase, and therefore prematurely born babies may survive. The process of alveolarization occurs when the existing airspaces subdivide as a result of the formation of new walls also called septa, and requires for its success the formation of a network of double-layered capillary at the base of the newly developing septum. The alveolarization process results in the formation of almost 90% of the gas exchange surface area. In addition, the network of double-layered capillary belonging to the immature septa starts to fuse, resulting in the formation of a single-layered and more optimized network for gas exchange.

The generation of branched tissues such as the lung requires tight control of the specification, initiation and elongation of the branch site. Recent research studies have led to the identification of many cellular and molecular mechanisms that regulate branching morphogenesis, vascular development and how multipotent stem and progenitor cell populations differentiate in the developing lung. However, more research is still needed to better understand these process that are crucial for lung morphogenesis, repair and regeneration and to identify therapeutic approaches for many respiratory disorders and congenital defects in humans.

Chapter 2

Advances in Lung Developmental Mechanobiology

Abstract

The lung contains both epithelial and mesenchymal cell types. Lung epithelial cells are characteristically localized at the interface between the organism and the environment and have many critical and vital functions such as the fluid balance, barrier protection, particulate clearance, production of both mucus and surfactants, and immune response initiation as well as tissue repair after injury. Lung cells are continuously exposed to mechanical stresses during their development and function. For example, lung epithelial cells are continuously exposed to varying levels of mechanical stresses due to lung's complex structure and the cyclic deformation of the lung during the respiratory cycle. The normal functions of the lung are maintained under these tightly regulated conditions, and changes in mechanical stresses may profoundly affect different functions of lung cells and therefore the overall lung functions. A major goal of lung mechanobiology is to understand how the mechanical behavior of the lung emerges from its cellular and molecular constituents. The central role of mechanics in the lung function was revealed with the help of the rapid progress in seminal historical developments, including both the identification and characterization of the functions of lung surfactants. In this chapter, we will describe the effects of mechanical factors on lung development, and how the airway peristalsis affects lung development. In addition, we will describe the functional roles of parathyroid hormone-related protein (PTHrP) in lung development and stretch transduction, as well as the functions of extracellular calcium-sensing receptor (CaSR) in fetal lung development.

39

Keywords: Lung; mechanobiology; mechanical factors; development; airway peristalsis; stretch transduction; parathyroid hormone-related protein (PTHrP); extracellular calcium-sensing receptor (CaSR).

2.1 Introduction

The main function of the lung is to facilitate the exchange of gases between the external environment and the circulation and this is linked with mechanics. On short timescales, every breath generates dynamic cycles of cell and matrix stretch, along with the convection of fluids in the vasculature and the airways. Perturbations such as shortening of airway smooth muscle (ASM) or rapid surfactant dysfunction alter respiratory mechanics with great influence on the lung function. On longer timescales, lung development, maturation, and remodeling strongly depend on cues from the mechanical environment. Thus, mechanics has long played a main role in developing our understanding of respiratory physiology and lung biology (reviewed in Warburton *et al.*, 2010; Morrisey and Hogan, 2010; Schittny, 2017).

Respiratory muscles can generate a trans-pulmonary pressure gradient, prompting gas to flow through the branched structure of the airways to the alveoli whose stability depends on a fine balance of tissue and surface forces (Fredberg and Kamm, 2006), while blood from the heart circulates through a network of capillaries to exchange O_2 and CO_2 between the alveolar–capillary walls (Maina and West, 2005). Understanding the physical origins of these functions, and their failure in some human diseases has been central to the study of medicine and respiratory physiology.

A major aim of lung mechanobiology is to understand how the mechanical behavior of the lung emerges from its cellular and molecular constituents (Suki and Bates, 2008). The central role of mechanics in lung function was revealed with the help of the rapid progress in seminal historical developments, including both the identification and characterization of the functions of lung surfactants (Obladen, 2005). In addition, at the macro scale, application of non-invasive imaging modalities facilitates the accurate measurements of the properties of intact lung mechanics locally and can also provide new information about defects and deformations of regional tissues. In intact lungs, for example, the microfocal X-ray imaging of lung airways can enable unprecedented measurements of airway dimensions and provide new insights into regional and axial as well as circumferential variations in airway strains, which occur with changing lung volumes (Sinclair *et al.*, 2007). In addition, magnetic resonance elastography (MRE) — a technique to sample tissue mechanics in soft organs — can also be used to measure the lung mechanics despite the fact that the application of MRE to the lung is complicated by its air-filled structure. Some data suggest that this methodology can be applied using porcine lungs inflated with hyperpolarized 3He to measure tissue mechanical properties within intact lungs (McGee *et al.*, 2008).

The cell-matrix model systems helped in identifying the stimuli that promote cell-mediated matrix remodeling, and concomitant changes in mechanical behavior and matrix organization (Leung *et al.*, 2007; Raub *et al.*, 2007, 2008). Several analyses of the second harmonic generation from multi-photon imaging of cell-matrix constructs have been used to explore linkages between the mechanical and structural properties of collagen matrices (Raub *et al.*, 2007, 2008). The mechanical testing of lung tissue strips and selective perturbation of matrix constituents have greatly advanced our understanding of the contributions of different cell-matrix protein types, such as elastin, collagen and other matrix proteins, to both tissue stability and deformability (Cavalcante *et al.*, 2005; Jesudason *et al.*, 2007). The functional roles of cell-matrix proteins in lung mechanics are further supported by studies on the lungs of knockout mice, such as genetically deficient mice for the proteoglycan decorin (Fust *et al.*, 2005). Knockout mice studies have also led to the assumption that lung mechanics can be partitioned into contributions from elastin at low volume to collagen at high volume. In addition, several studies have led to a hypothesis explaining the progressive emphysema on the basis of percolation of sequential alveolar wall rupture based on several evidences showing that the alveolar walls can fail under loading, particularly in the remodeled matrix (Ito *et al.*, 2006; Ritter *et al.*, 2009; Bates *et al.*, 2007; Suki and Bates, 2008). These advanced approaches are improving our understanding of lung tissue mechanics and can lead to the determination of how lung microme-chanics are changed by specific molecular perturbations.

Furthermore, the biofluid mechanics in the respiratory system are the subject of several studies that can lead to a remarkable improve-ment in our understanding of the mechanisms of lung injury that are associated with the opening and closure of fragile tissue structures in parallel with current advances in the field of solid mechanics (Bertram and Gaver, 2005; Yalcin *et al.*, 2007). This leads to the generation of multi-scale models for the study of fluid dynamics in both lung air-ways and vasculature (Tawhai and Burrowes, 2008). Interestingly, the airway-lining layer of mucus can obviously respond to airway shear stresses at the interface of fluid and solid surfaces (Tarran *et al.*, 2005, 2006). Moreover, the remarkable pathological changes in the viscoe-lasticity of mucus are considered as a new therapeutic modality for cystic fibrosis patients (Donaldson *et al.*, 2006; Tarran *et al.*, 2007).

The dysregulation of the dimensions of airways is an important pathogenetic mechanism in some severe pulmonary diseases, such as chronic obstructive pulmonary disease (COPD) and asthma, because the narrowing of airways can severely affect the efficiency of gas trans-port. Indeed, flow resistance through the airways was reported to be highly responsive to the airway dimensions that can be dynamically regulated by the contractions of the ASMs. In addition, great advances in the field of lung mechanobiology have been achieved with the finding that the tone of ASMs is dynamically equilibrated, and consequently is sensitive to lung volume history (An *et al.*, 2007). Better understanding of the relevant physics of these processes can help to identify important factors that govern lung physiology, devel-opment, repair and regeneration.

Growth factor-regulated growth requires transport of material driven by local pressures. Polymers of the extracellular matrix (ECM) not only bind to growth factors, but also provide resistance to deformation, diffusion and fluid flow, and transmit force over long distances. Cytoskeletal elements, in turn, generate and transmit forces and conversely provide resistance to deformation, while adhesion mol-ecules regulate both tissue structure and cell motility, and these in turn interact through deformations and forces (Warburton *et al.*, 2010).

An embryonic tissue is mechanically like a viscoelastic fluid (Forgacs *et al.*, 1998; Jakab *et al.*, 2008a, 2008b). Thus, on the timescale of growth, the elastic component of tissue viscoelasticity may be neglected, and the tissue behaves mechanically as a fluid, so that a simple mechanobiological model (Lubkin and Murray, 1995) can predict the relationship between morphology and pressure that is observed in lungs with occluded tracheal fluid outflow (Unbekandt *et al.*, 2008).

In developing tissues, differentiation is very sensitive to ECM mechanics. For example, mesenchymal stem cells *in vitro* differentiate towards neurons at 1 kPa, towards muscle at 10 kPa, and towards cartilage at 30 kPa (Engler *et al.*, 2006; Warburton *et al.*, 2010).

The model of the mechanobiology of embryonic lung morphogenesis has received more attention (reviewed by Warburton *et al.*, 2010). Hypothesizing that a tissue regulates itself and its environment in order to maintain an equilibrium level of tangential stress on an epithelium, the model of the mechanobiology of pseudoglandular lung morphogenesis (Lubkin and Murray, 1995) treated the epithelium mechanically as a viscous fluid with a surface tension (Foty *et al.*, 1994). The model predicts that the size of the branches generated by the epithelium should be inversely related to the pressure difference between the lumen and the external medium, which is the native mesenchyme. There is proof that embryonic lung epithelium regulates its tangential stress by regulating cytoskeletal tension via the Rho-ROCK system (Moore *et al.*, 2005).

2.2 Effects of Mechanical Factors on Lung Development

Pulmonary hypoplasia secondary to oligohydraminos and congenital diaphragmatic hernia (CDH), etc., is an important cause of neonatal morbidity and mortality. In fact, pulmonary hypoplasia is the most common finding (up to 26%) in neonatal autopsies (Husain and Hessel, 1993). Moreover, more than 20,000 babies are born every year in the United States before 27 weeks of gestation (canalicular stage of lung development). These disorders have in common an incomplete development of the lungs. In addition to the risk of death, these conditions can also cause severe respiratory distress at birth and serious long-term morbidities (Liu and Post, 2000; Wilson-Costello, 2005). Currently, the management of these conditions is primarily supportive in nature and there is no specific way to accelerate the development of the lungs.

Lungs are unique in that their growth and development depends primarily on extrinsic factors and specifically on mechanical forces (Liu and Post, 2000; Sanchez-Esteban, 2001). During gestation, the epithelium of the lung secretes fluid creating a constant distension pressure in the lumen of the lung of approximately 2.5 mmHg (Scarpelli, 1975). Moreover, the fetus makes episodic breathing movements (FBMs) starting in the first trimester and increasing in frequency up to 30% of the time by birth (Harding, 1997). Whereas there is agreement on how much pressure is generated inside the fetal lung by fluid secretion, the same does not apply to the change in length experienced by the lung with each FBM. Part of this controversy is due to the complex and variable changes in thoracic dimensions and intraluminal pressures generated by FBM (Harding, 1996). During non-accentuated periods of FBM, the intraluminal pressures may only decrease by 2–3 mmHg. However, this pressure can decrease up to 10–15 mmHg during accentuated periods of FBM (Polglase, 2004).

It has been suggested that distension of the fetal lung generated by FBM is negligible, as they generate very little tidal movement of fluids (Hooper and Wallace, 2006). Based on changes in the thoracic shape observed during FBM and assuming the spherical shape of the distal potential airspaces and the cone shape of the thorax, it has been speculated that FBM might result in repetitive changes in distal lung surface area of about 5% (Kitterman, 1996). Fetal type II epithelial cells have shown that a stimulus of that magnitude results in cell proliferation (Liu, 1999) and differentiation (Sanchez-Esteban, 2001). In either case, it is clear from experimental animals that abolition of FBM (Goldstein and Reid, 1980) or drainage of lung fluid volume (Moessinger, 1990) lead to lung hypoplasia. Therefore, both cyclic mechanical deformation and tonic hydrostatic distension provide physical signals necessary for fetal lung development. However, the mechanisms by which the fetal lung senses these mechanical signals to promote development are not well characterized.

Previous investigations in fetal lambs have shown that lung fluid composition after tracheal ligation was critical to promote lung development as acceleration of growth and differentiation was not observed when lung fluids were replaced with normal saline (Luks *et al.*, 2001). These studies suggest that increased intra-tracheal pressure after tracheal ligation releases soluble factors that are important for lung development. This hypothesis is supported by previous laboratory-based *in vitro* studies in which fetal type II epithelial cells isolated during the canalicular stage of lung development were exposed to mechanical strain mimicking mechanical forces in lung development. Sanchez-Esteban (2013) showed that mechanical strain cleavages and releases the soluble mature forms of epidermal growth factor receptor (EGFR) ligands HB-EGF and TGF-a (Huang *et al.*, 2012; Wang *et al.*, 2009). Release of these soluble factors bind and activate the EGFR via autocrine or paracrine signaling and promote differentiation of type II cells via the ERK signaling pathway (Wang *et al.*, 2013).

The identification of soluble factors released by mechanical forces that are important for normal lung development could lead to new avenues to accelerate lung development. Potential translational research applications would be prenatal administration to fetuses affected by pulmonary hypoplasia secondary to oligohydramnios or CDH or fetuses with borderline viability (22–24 weeks) and at risk for delivery.

Another theoretical application would be post-natal administration through the endotracheal tube. This is just an example of how the information obtained from these *in vitro* mechanistic studies could have the potential for clinical applicability. However, before considering their use in humans, rigorous experiments in animal models are required to demonstrate the effectiveness of these therapies.

Further examples on the intimate relation between lung development and mechanobiology, include the impact of intrathoracic herniation of abdominal viscera on lung growth in CDH and the role of iatrogenic barotrauma on preterm lung in the genesis of broncho-pulmonary dysplasia (BPD) (Harding and Hooper, 1996).

Several human birth defects and *in utero* experiments have demonstrated how lung development requires the influence of mechanical factors. Most frequent among them is the CDH that is characterized by a remarkable diaphragmatic defect and lung hypoplasia as well as an intrathoracic herniation of abdominal viscera (Smith *et al.*, 2005).

Newborns affected by CDH have a high mortality rate principally due to inadequate lung growth. The failure of normal lung development has been attributed to mechanical compression of the lung by the herniated abdominal viscera (Starrett and de Lorimier, 1975). In infants with CDH, the lungs have reduced weight and volumes with histologic features of developmental immaturity, manifested by a reduction in the size and number of alveoli. Several mechanisms have been hypothesized to cause CDH-associated lung hypoplasia. Classic teaching indicated that the compression of the lung by herniated viscera and the lack of respiratory movements were the causes of hypoplasia phenotype; however, other studies suggest that abnormal lung development precedes the herniation of abdominal contents into the thoracic cavity, and it is therefore a primary defect. These observations have led to the conclusion that lung defects are at least partially independent of the diaphragm defect (Jesudason *et al.*, 2000; Keijer *et al.*, 2000).

Similarly, human fetuses with profound renal failure and/or renal agenesis exhibit the Potter's syndrome phenotype. In this situation, lung hypoplasia is theorized to occur due to excessive lung fluid loss to the underfilled amniotic cavity and/or extrinsic compression of the fetal thorax owing to the same lack of amniotic fluid. Support of this thesis emerges from ovine fetal models where bilateral nephrectomy impairs lung growth (Wilson *et al.*, 1993).

A competing view is that much of the lung hypoplasia associated with both renal agenesis and CDH results from early developmental insults to the lung that may precede or coincide with the genesis of the renal agenesis and diaphragmatic defect, respectively. Support for this view emanates from the teratogenic nitrofen rodent model of the CDH wherein early lung malformation precedes CDH development

(Jesudason *et al.*, 2000). Similarly, a transgenic murine model of renal dysgenesis has demonstrated that lung hypoplasia emerges prior to fetal urine output making a significant contribution to the amniotic fluid (Smith *et al.*, 2006).

In addition to the extrinsic forces acting upon the fetal lung, there is a distending pressure within the normal developing lung that is due to the production of lung liquid. Drainage of this fluid and loss of this pressure that is generated by fetal tracheostomy is associated with lung hypoplasia (Fewell *et al.*, 1983). Conversely, retention of this fluid in human fetuses with congenital laryngeal atresia is associated with lung distension and overgrowth (Harding and Hooper, 1996).

Transgenic inhibition of skeletal muscle formation is also associated with lung hypoplasia. For example, the diaphragm muscle in MyoD−/− mice is significantly thinned and cannot support fetal breathing movements. As a result, lung is hypoplastic, and the number of proliferating lung cells is decreased in MyoD−/− lungs at E18.5 (Inanlou and Kablar, 2003). Therefore, at this stage, mechanical forces generated by the contractile activity of the diaphragm muscle play an important role in normal lung growth (Inanlou and Kablar, 2003). Post-natally, in premature newborns, the mechanical effects of positive pressure ventilation collaborate with other factors as inflammation to generate BPD (Warburton *et al.*, 2001). Beyond the newborn period, the influence of mechanical factors appears to persist. Following lung resection, there is the compensatory lung growth so there is a mixture of lung distension and parenchymal growth (Thurlbeck, 1983a, 1983b). This compensatory response to pneumonectomy suggests a capacity for the lung to respond to a changed mechanical environment and to alteration in the respiratory surface area for gas exchange.

ASM hypertrophy and hyperreactivity in asthma is associated with air trapping and acute lung distension; however, with time, this is associated with airway remodeling and chronic lung hyperexpansion. In asthma, ASM-led airway occlusions may therefore have analogous effects on fetal tracheal occlusion, which remodels and distends prenatal lung (Jesudason, 2007). Moreover, transient endogenous ASM-led airway occlusions occur in fetal lung known as airway peristalsis. This contractility may be a regulator for lung growth (Jesudason, 2006). With this published information taken into consideration, Section 2.4 focuses on three areas of current interest in lung mechanobiology such as lung liquid secretion, airway peristalsis and the role of calcium signaling in this secretory and contractile environment.

2.3 Effects of Airway Peristalsis on Lung Development

Airway peristalsis (AP) produces spontaneous phasic airway contractions and transient, reversible airway occlusions throughout normal lung development (Jesudason *et al.*, 2005; Pandya *et al.*, 2006; Parvez *et al.*, 2006). AP begins as soon as the smooth muscle (SM) develops, with pseudoglandular branching, and becomes more vigorous towards later stages of development. AP has been shown to influence lung development and its blockage affects lung development (Jesudason, 2009) due to unclear mechanisms. Jesudason (2009) hypothesized that AP critically modulates physical forces on airway cells, influencing the patterning of branching morphogenesis and, ultimately, the overall growth of the lung (Jesudason, 2006; Warburton and Olver, 1997). A more specific hypothesis would consider that AP involves fluid flow and multiple tissues that operate differently in these different contexts.

AP stimulates growth so its absence retards growth. It was hypothesized that mechanical stimulation from AP promotes growth (Jesudason, 2009). Tracheal closure enhances branching (Unbekandt *et al.*, 2008) and, therefore, Bokkaa *et al.* (2015) hypothesized that the mechanical stimulation from lumen pressure promotes branching. These hypotheses arise from experiments at the organ level, where it is hard to identify the specific mechanisms by which the mechanical inputs alter morphogenesis. At a minimum, mechanical stimulation can be expected to affect different locations differently, and that will affect different tissues differently. Furthermore, it is known that many *in vitro* systems show enhanced growth with mechanical stimulation.

AP compresses the epithelium in the stalk circumferentially as it is elongated apicobasally. This transient compression may play a role in organizing the cellular arrangement in the tissue, which could affect the extension of the tubule (Bokkaa *et al.*, 2015).

Tracheal occlusion enhances branching, presumably stimulating tissues through increased lumen pressure (Unbekandt *et al.*, 2008). AP enhances lung growth, and it has been hypothesized that this too is primarily a pressure effect. However, an alternative hypothesis is that the cellular stimulus is due to the stretch of a specific area and not due to the pressure. In addition, viscosity of the lumen fluid affects the mechanics of AP only if the trachea is open and fluid can drain and refill during AP. For an open trachea, as it occurs *in vivo*, the more viscous the fluid, the greater the pressure from AP, and therefore the greater the stretch of the tip from AP. Thus, there may be an interaction between AP, mechanical stimulation of growth and/or branching, and developmental disorders involving mucus production.

The embryonic lung is filled with liquid. Active chloride secretion through the CFTR and other chloride ion channels attracts sodium and thus creates an ionic and osmotic gradient, which draws water into the lumen. Lung liquid is produced starting from the earliest stages of embryonic lung development up to delivery. The hydraulic pressure within the lung lumen is determined by the rate of production of liquid together with the pinch-cock valve function of the primitive larynx.

Obstruction of fluid outflow by clipping or cauterization of the embryonic trachea increases the intraluminal pressure by about 2- to 3-fold. This is accompanied by a 3-fold increase in the branching of the embryonic airway. Using real-time microscopic cinematography, stereotypic airway branching can be parsed temporally into a branch extension phase, an arrest phase and a budding phase.

Once the budding phase is complete, the branch extension follows once more. Following tracheal occlusion, the rate of bud extension increases by about 2-fold while the inter-bud distance is about halved. These effects of increased intraluminal pressure depend upon fibroblast growth factor-10 (FGF10)-FGFR2b-Sprouty signaling (Unbekandt *et al.*, 2008).

Airway peristalsis is the spontaneous ASM contractility that occurs in species as diverse as birds and humans and which increases in frequency from embryonic stages through to birth (Schittny *et al.*, 2000). Propagating intercellular ASM calcium waves precede each peristaltic contraction (Featherstone *et al.*, 2005). These waves of ASM activity emanate from pacemaker areas in the proximal airway before being transmitted to the distal airways (Jesudason *et al.*, 2005).

In asthma, this pacemaker-derived rhythm may even be important postnatally (Jesudason *et al.*, 2006); hence the pacemakers may be a target for ablation by bronchial thermoplasty (Jesudason, 2009). Prenatally, the peristaltic contractions appear to induce rhythmic airway occlusions that direct a pulsatile wave of lung fluid preferentially toward the lung's growing tips. This phasic activity results in the relaxation of the growing lung buds and a rhythmic stretch. Hence, airway peristalsis may provide an endogenous form of airway occlusion and may regulate both stretch and pressure in the growth tips of the developing lung (Jesudason, 2009).

2.4 Roles of Parathyroid Hormone-related Protein (PTHrP) in Lung Development and Stretch Transduction

Parathyroid hormone-related protein (PTHrP) is expressed in normal and malignant lungs and has roles in development, homeostasis, and pathophysiology of injury and cancer. Effect of PTHrP in the developing lung include type II cell maturation and regulation of lung branching morphogenesis. In the *adult lung*, PTHrP inhibits type II cell growth, stimulates disaturated phosphatidylcholine secretion, and sensitizes them to apoptosis. In lung cancer, PTHrP may play a role in carcinoma progression or metastasis. The protein could be a useful marker for assessing type II cell function or lung maturity, predicting risk of injury, and detecting lung cancer. PTHrP-based therapies could also prove useful in lung cancer and lung injury (Hastings, 2004). In addition, when alveolar type II cells are stretched, PTHrP is produced (Torday *et al.*, 2002). PTHrP binds in a paracrine manner to its cognate receptor on adjacent adepithelial fibroblasts (Torday and Rehan, 2002), inducing these to differentiate into a lipofibroblast lineage (Schultz *et al.*, 2002).

The lipofibroblasts, first discovered by Vaccaro and Brody (1978), serve a dual purpose in the alveolar interstitium, producing leptin and protecting the lung against oxidant injury (Torday *et al.*, 2001), which binds in a converse paracrine loop to its cognate receptor on the alveolar type II cell, thus stimulating the synthesis of protein and surfactant phospholipids (Torday *et al.*, 2002a, 2002b). PTHrP stimulates the vasodilation of the alveolar capillaries (Gao and Raj, 2005), thereby coordinating surfactant production with increased alveolar perfusion in response to alveolar stretching, a dynamic process referred to by lung physiologists as ventilation-perfusion matching.

Fluid distension also induces sonic hedgehoc (Shh) expression by the endodermal cells, which in turn stimulates the Wingless/int (Wnt) signaling pathway in the surrounding mesoderm. Wnt in turn drives increased PTHrP expression by the endoderm, which then feeds back to negatively regulate the Wnt pathway by stimulating Protein Kinase A. In addition, PTHrP stimulates fibroblasts to differentiate into a lipofibroblast lineage. Moreover, Wnt/beta catenin signaling has recently been functionally linked to both the CFTR and PTHrP (Cohen *et al.*, 2008).

2.5 The Role of Extracellular Calcium-Sensing Receptor (CaSR) in Fetal Lung Development

In fetal mammals, serum levels of both total and ionized calcium significantly exceed those in adults. This relative fetal hypercalcemia is crucial for skeletal development and is maintained irrespective of maternal serum calcium levels. CaSR has a role in creating and maintaining this relative fetal hypercalcemia through the regulation of parathyroid hormone-related peptide secretion (Riccardi *et al.*, 2013).

Optimal postnatal lung function is dependent on a tightly controlled embryonic lung development program. The program has been divided into five stages: embryonic; pseudoglandular; canalicular; terminal saccular and alveolar (Warburton *et al.*, 2010; Whitsett *et al.*, 2004).

Branching morphogenesis, which occurs mainly in the pseudoglandular stage, is an important part of lung development. Despite the apparent complexities of lung branching, the budding patterns at early stages take place through only three simple genetically-encoded modes of branching termed as domain-branching, planar bifurcation, and orthogonal bifurcation which are continuously repeated to form various branching compositions (Metzger, 2008).

Several studies have identified a role for Ca2+ in lung development. Roman (1995) demonstrated that the L-type calcium channel blocker, nifedipine, abolished spontaneous airway contractions and impaired branching morphogenesis indicating that Ca2+ influx is necessary for normal lung development. Furthermore, work by Jesudason *et al.* (2005) has confirmed these observations and demonstrated that nifedipine abolishes both lung growth and peristaltic SM contractions mediated by FGF-10, a critical intrinsic factor in mammalian lung development (Park *et al.*, 1998).

Peristaltic waves initiated by pacemaker areas within the proximal airway can regulate intraluminal pressure through rhythmic phases of both relaxation and stretching. In addition, they can produce fluid pulses that are directed to the distal buds of the lung (Jesudason *et al.*, 2005; Jesudason, 2009). The frequency of these waves is positively associated with lung growth rates (Jesudason, 2005). Measurements of intracellular Ca2+ using Ca2+-sensitive fluorescent dyes have revealed that contractions are induced by regenerative source waves in ASM cells that are dependent on both extra- and intra-cellular Ca2+ stores (Featherstone *et al.*, 2005). However, little is still known regarding the role of Ca2+o as a signal in lung development, or whether the CaSR, a key mediator of the actions of Ca2+o, was expressed in the fetal lung.

Initial observations revealed that the CaSR is widely expressed in developing human and mouse fetal lungs, particularly in the pseudoglandular stage (Finney, 2008). These studies have shown that lung branching is sensitive to Ca2+o over the range 0.5–2.0 mM. Interestingly, higher Ca2+o levels, like those seen in fetal blood (~1.70 mM), had a suppressive effect on branching over 48 h. Furthermore, the effect was mimicked in lungs exposed to the calcimimetic NPS R-568 in the presence of 1.05 mM Ca2+o, demonstrating that CaSR activation negatively modulates branching morphogenesis (Finney *et al.*, 2008). Other studies using the mouse lung explant culture model have demonstrated that fetal [Ca2+]o represents an important extrinsic controller of fetal lung development by suppressing branching morphogenesis and cellular proliferation while enhancing Cl-dependent fluid secretion (Finney *et al.*, 2008). Since all the effects of fetal [Ca2+]o can be mimicked by the pharmacological activation of the G protein-coupled, extracellular Ca2+-sensing receptor, CaSR, it appears that the fetal Ca2+, acting through this receptor, is an important extrinsic factor, which allows optimal lung development by balancing branching morphogenesis with lumen distension.

The CaSR is expressed exclusively in the epithelium of the embryonic mouse lung at the beginning of the pseudoglandular phase, and its expression is maintained in the organ explant culture model for at least 48 h. From E14.5 in the mouse (and week 9 in the human fetal lung), the CaSR protein expression can also be detected in the mesenchyme. At the inception of the canalicular phase, the CaSR expression begins to wane and is completely absent from the adult rat, mouse and bovine lung (Finney *et al.*, 2008; Riccardi *et al.*, 1995; Brown *et al.*, 1993). CaSR-dependent regulation of both branching morphogenesis and fluid secretion occurs through phospholipase C-dependent increase in intracellular Ca2+ concentration and phosphoinositide 3(PI3) kinase activation.

Whether CaSR-dependent increases in intracellular Ca2+ directly affect lung development via the formation of peristaltic waves and/or control lung growth by altering epithelial membrane integrity through activation of PI3 kinase-dependent beta-catenin signaling continues to be under investigation in different research laboratories.

2.6   Summary and Conclusions

The lung has a highly complex and dynamic structure that allows more efficient gas exchange. The application of mechanical forces is required for gas exchange process since it acts to distend lung structures and prevent the collapse of pre-stressed lung structures or units. This dynamic and effective dynamic mechanical environment functions to maintain the proper physiological lung functions.

Recent studies indicate that remarkable changes in the mechanical properties of lung tissues or the applied mechanical forces on these tissues can be caused by or contribute to different types of lung injuries or diseases. This is well studied in lung epithelial cells that exist at the interface between the organism and the surrounding environment and are sensitive to such changes in the lung tissue properties or deforming stress. In addition, many recent studies have described different types of strains and stresses in the lung and their ways of transmission as well as different types of approaches that have been developed to determine how lung cells respond to mechanical stresses. Studies have also focused on how various types of stresses and strains may vary in a wide range of human diseases or disorders and animal models. More studies are needed to better understand the sense and response of lung cells to changes in mechanical strains of stress since this will help in identifying the functional roles of different types of lung cells, including epithelial cells, in the normal development, morphogenesis and function of the lung. It will also help in understanding how the functions of lung epithelial cells can change in lung diseases, including asthma, fibrosis, emphysema, and acute lung injury.

Mechanical forces can play important roles in normal lung development, morphogenesis and function, as well as in many pathological changes associated with lung diseases. Different types of mechanical stresses that take place during normal breathing and in pathological changes in the lung are relatively well characterized. However, more research is still needed to answer questions about the distribution and magnitude as well as the heterogeneity of these mechanical forces *in vivo*.

Despite their current challenges, future research and clinical studies that aim to identify, characterize, measure and localize the mechanical forces affecting the lung will help improving our understanding of functions of the lung and how these functions change in lung disorders or diseases.

Epithelial cells have a unique localization at the interface, which is subjected to mechanical stress changes continuously. The proliferation and differentiation of lung epithelial cells, and, the fetal breath movements and alterations in lung mechanics after birth affect epithelial cell proliferation and differentiation in the lung, but the underlying mechanisms controlling these processes are currently not well understood and still need more investigations. Furthermore, there are several changes in mechanical forces after acute injury of the lung or in the lung developing fibrosis, asthma, emphysema, or other pathological conditions. These changes can result in epithelial cell injury or alter epithelial cell differentiation. Yet, the underlying mechanisms that can lead to lung injury or cell phenotype changes in response to alterations of the mechanical forces are currently not clear.

Rapid advances have occurred in developing new approaches to uncover cell sensing and responses to different types of mechanical stresses. These advances will also lead to developing more effective approaches to investigate the responses of lung epithelial cells in culture, and to study the mechanisms underlying the responses to mechanical stresses in intact animals or tissues more effectively in the future.

Chapter 3

Pulmonary Complications Associated with Preterm Birth

Abstract

Preterm birth results in several respiratory complications, one of the major causes of infant morbidity and mortality. There is a high risk of development of chronic lung disease and respiratory distress syndrome among preterm infants. In addition, follow-up studies show that the functional impairment of airways is commonly seen in preterm children. Bronchopulmonary dysplasia (BPD) is one form of *chronic lung disease* affecting most premature newborns and infants, and results from lung damages. Studies on surviving children with BPD suggest persisting defects in the structure of both the airways and the parenchyma of the lung. Preterm birth survivors have several respiratory consequences such as immediate breathing challenges due to lungs that are still underdeveloped, which usually manifest as respiratory distress syndrome (RDS). The current therapies for neonatal lung diseases enable the survival of preterm infants who are born at week 22 of pregnancy or later, although they still suffer from some health consequences. Stem cells have therapeutic applications and potential for the treatment of infants with pulmonary complications associated with preterm birth, such as BPD.

Keywords: Lung; preterm; chronic lung disease; respiratory distress syndrome; broncho-pulmonary dysplasia; mesenchymal stem cells; bone marrow.

3.1 Major Pulmonary Complications Associated with Preterm Birth

With preterm birth, respiratory failure is a common morbidity because of the structural and physiological immaturity of the lung. The lung is occupied by fluid during fetal development. This fluid is secreted by lung epithelial cell types and is critical in promoting the growth of the lung (Harding and Hooper, 1996). The lung normally 6 switches from fluid secretion to fluid absorption just before birth. Mature alveolar epithelial cells are able to achieve this alveolar fluid clearance (AFC) by unidirectional Na^+ transport by means of the epithelial Na^+ channels (ENaC) and the Na and K-ATPase. This creates a driving force for absorption of fluid. In addition, mature alveolar type II (ATII) cells produce surfactant into the lumen of the lung. This reduces the collapse of the lung during end exhalation by decreasing the surface tension within the alveoli and terminal airways (Walsh *et al.*, 2013a, 2013b). Surfactant is formed by neutral lipids, phospholipids, and proteins, and this forms a film between the alveolar gas and the terminal airways/ alveolar surfaces. In addition, surfactant reduces lung inflammation and enhances mucociliary clearance (Walsh *et al.*, 2013). Fluid accumulation and alveolar instability is due to impaired AFC because of insufficient expression of Na^+ channels (O'Brodovich, 1996), and the structural immaturity of the lung as well as surfactant deficiency with low alveolar numbers which leads to impairment of the function of the lung.

Gestational age is inversely correlated with the incidence of neonatal respiratory distress syndrome (RDS). This results in an RDS risk of more than 50% for infants weighing <1000 g and born before 30 weeks of gestational period (Ramanathan, 2006). The respiratory outcome has been shown to be improved by some therapeutic options which include antenatal glucocorticoid treatment and significant surfactant replacement therapy for the compensation of surfactant deficiency.

Antenatal glucocorticoids in low doses speeds the late-gestation maturation of the lung through increasing the ATII cells volume density, enhancing surfactant synthesis, and upregulating AFC (Ballard *et al.*, 1997, Folkesson *et al.*, 2000; Snyder *et al.*, 1992). It is routinely given to women who are at risk of preterm birth to help in improving lung function and perinatal survival (Roberts and Dalziel, 2006). Furthermore, glucocorticoids may be prescribed for improving lung functions and gradually deprive it from postnatal mechanical ventilation. Glucocorticoids have been shown to decrease alveolar crest formation in animal experiments and may, therefore, worsen alveolar simplification (Tschanz *et al.*, 1995). Furthermore, post-natal administration of glucocorticoids may have a negative effect since it can be associated with some long-term and severe neuro-developmental side effects (Yeh *et al.*, 2004).

Both oxygen supplementation and mechanical ventilation are needed to provide sufficient oxygen supply, despite the different types of clinical interventions currently available. However, the lung lacks both anti-inflammatory mediators and antioxidant capacity, which leads to enhanced oxygen toxicity and inflammatory processes, and finally lung tissue destruction. Insufficient repair results in persistently low gas exchange capacity and simplification of the alveoli which is the sign of bronchopulmonary dysplasia (BPD). Incidence of BPD/chronic lung disease of prematurity is correlated with the increased survival rate of infants with very-low-birth-weight who are recovered from RDS (Walsh *et al.*, 2006). This affects 30% of preterm infants born before 29 weeks of gestational age (Smith *et al.*, 2005). BPD, therefore, represents a problem happening from complications related to abnormal development of lung that occurs in older infants or due to RDS treatment (Antunes *et al.*, 2014).

The phenotype of BPD has changed over time. The "old" BPD was characterized by lung edema, airway epithelial metaplasia, pronounced inflammation, severe lung injury, marked airway and pulmonary vascular smooth muscle hypertrophy, and peribronchial fibrosis (Hilgendorff *et al.*, 2014; Northway *et al.*, 1967). In contrast, the "new" BPD differs from the "old" form as it has less structural damage due to treatment with surfactant and antenatal glucocorticoids and optimized ventilation strategies. Unlike the "old" BDP, the "new" BPD displays only mild peribronchial fibrosis and inflammation, while structural immaturity of the lung predominates (Baraldi and Filippone, 2007; Jobe, 1999).

Histologically, the "new" BPD is characterized by thickened alveolar septa, alveolar hypoplasia (fewer and larger alveoli), mild hypertrophy of both vascular smooth muscles, and airway abnormal deposition of extracellular matrix components, dysmorphic pulmonary microvascular networks, interstitial fluid accumulation, and an arrest of lung development at the late canalicular to early saccular stage (Coalson *et al.*, 1999; Thibeault *et al.*, 2003a, 2003b). These conditions cause chronic pulmonary malfunction with significant morbidity and mortality (Baraldi and Filippone, 2007). Lung developmental arrest and damage due to BPD are unrepairable, and the respiratory dysfnction may continue into adult life (Gross *et al.*, 1998; Northway *et al.*, 1990). Currently, BPD therapy is mainly soothing and limited to moderately active substances such as high caloric feeding, caffeine, and glucocorticoids or vitamin A (Baveja and Christou, 2006). Thus, a therapeutic strategy would be highly-desired to restart alveolar growth and promote the development of normal and complex alveolar structures.

3.2 Therapeutic Potential of Mesenchymal Stem Cells (MSCs) in Lung Injury

Preterm infants usually suffer from complications in the lung that lead to significant morbidity and mortality. Physiological and structural and lung immaturity impairs perinatal lung transition to air breathing which causes respiratory distress. Both mechanical ventilation and oxygen supplementation provide sufficient oxygen supply. However, they enhance the inflammatory processes which might lead to a chronic lung disease called BPD. Current therapeutics that prevent or treat BPD are limited and have serious side effects, highlighting the need for new therapeutic approaches such as the mesenchymal stem cells (MSCs) that have provided therapeutic potential in animal models of BPD.

Multiple animal studies suggest that MSCs may have therapeutic potential for infants with BPD, either to treat infants with established BPD or to prevent BPD (Aslam *et al.*, 2009). Newborn rodents are used as BPD models since rats and mice are born in the saccular stage of lung development, which overlaps with the human preterm infant between the 24th and 28th week of gestation although newborn rodents do not usually suffer from respiratory distress. Therefore, rodent pups should be exposed to high concentrations of oxygen at 60–95% O_2 to produce lung injury (Warner *et al.*, 1998). Lung injury in neonatal rodents exposed to hyperoxia for 10–14 days shares few pathological features of BPD, including pulmonary edema, fibroblast proliferation, inflammatory cell infiltration, collagen deposition, and structural disruption affecting vascularization and alveolarization (Hilgendorff *et al.*, 2014).

Many research studies have addressed the potential therapeutic applications of MSCs in lung injury induced by hypoxia. Bone marrow (BM)-MSC transplantation can successfully reduce lung inflammation, as evidenced by a decrease of macrophage and neutrophil number in the bronchoalveolar lavage fluid of newborn murine pups that were assessed two weeks after hyperoxia exposure (Aslam *et al.*, 2009). Remarkably, transplanted BM-MSCs can enhance pulmonary vascular density and improve survival, alveolarization, and exercise capacity (van Haaften *et al.*, 2009). These striking improvements were due to BM-MSC delivery on day 4 of hyperoxia, whereas application on day 14 did not improve lung architecture, suggesting a preventive rather than a regenerative effect of MSCs (van Haaften *et al.*, 2009).

In the developing lung, loss of the populations of stem and progenitor cells may be the cause of oxygen-induced depletion of lung resident endothelial progenitor or angiogenic cells and tissue simplification. That may be a possible mechanism for the arrested lung vascular growth in BPD (Balasubramaniam *et al.*, 2007; van Haaften *et al.*, 2009). This is consistent with the significant reduction of both endothelial progenitors and angiogenic cells in lungs of hyperoxia-exposed newborn mice (Balasubramaniam *et al.*, 2007) and in samples of cord blood from premature infants who can develop BPD

(Baker *et al.*, 2012). In addition, BM-MSC transplantation can lead to an increase of the resident bronchoalveolar stem cells, which are a lung cell type that may contribute to alveolar repair after injury (Tropea *et al.*, 2012). MSCs can, therefore, represent a potential and effective therapeutic option for the treatment of BPD since they can replace the depleted endothelial stem/progenitor cells or restore their functions (Alphonse *et al.*, 2012; Anversa *et al.*, 2012).

3.3 Summary and Conclusions

A tightly regulated series of highly orchestrated events are needed for proper development of the lung. Preterm birth refers to babies who are born before 37 completed weeks of pregnancy. Different types of pre- and post-natal factors and mechanisms that are associated with preterm birth can compromise the proper lung development. Preterm birth is associated with several respiratory complications that are one of the major factors leading to infant morbidity and mortality. Thus, there is a high risk of developing severe lung diseases, including respiratory distress syndrome and BPD that are forms of chronic lung disease, among preterm infants, leading to lung damages. Surviving children with BPD may have persisting abnormalities in the structure of lung airways and parenchyma. Preterm birth survivors have also several other respiratory consequences, including RDS.

The available therapies for neonatal lung diseases can currently permit the survival of preterm infants who are born at week 22 of pregnancy or later. However, these preterm infants have major health consequences. Consequently, there is a need for more effective therapeutic approaches for the treatments of respiratory complications that are associated with preterm birth. Recent studies have focused on the potential therapeutic applications of stem cells in the treatment of lung injury and pulmonary complications associated with preterm birth such as BPD. Indeed, there are evidences that mesenchymal stem cells (MSCs) may act as a potential and effective therapeutic option for the treatment of BPD since these stem cells can replace the depleted endothelial stem/progenitor cells or restore their functions. However, more research is still needed to investigate the functions and applications of different stem cell types in the treatment of pulmonary complications related to preterm birth.

Furthermore, there is a critical need to determine and characterize the molecular and cellular mechanisms by which the development of lung is abnormally changed in preterm newborns and to improve the strategies of both managing the preterm labor and providing essential support to preterm newborns, both of which have a low impact on the lung growth.

Chapter 4

Lung Stem and Progenitor Cells

Abstract

Both the conducting airways (the bronchiolar and tracheobronchial airways) and the gas exchange airspaces (the alveolar regions) are the two major distinct compartments of the lung. Each of these regions contains distinct types of epithelial cells with characteristic compartments of stem and progenitor cells that include epithelial and alveolar stem/progenitor cells, stem and progenitor cells of the bronchi and trachea, and lung mesenchymal stem and progenitor cells such as airway smooth muscle (ASM) stem and progenitor cells and vascular stem and progenitor cells. In addition, the cellular plasticity of different lung-specific stem cells is currently an emerging field of research. Many recent studies have, therefore, used lineage tracing to determine distinct populations of epithelial stem and progenitor cells in the lung. Moreover, stem cells are likely involved in some major lung diseases. Yet, little is known about the functions of different stem cell types in the lung, despite common agreements that lung stem cells may have a major role in the repair and regeneration of the lung. In this chapter, we describe different types of stem and progenitor cells in the lung and mechanisms that regulate both their development, including their proliferation and differentiation, and plasticity during lung morphogenesis, as well as stem cell contribution to lung repair and regeneration. Furthermore, we describe stem cell-related diseases in the lung and stem cell contributions to both the immunomodulation of lung diseases and lung repair and regeneration, as well as the potential of stem cell-based therapies in different lung diseases.

Keywords: Lung; alveolar stem/progenitor cells; bronchial and tracheal stem/progenitor cells; lung mesenchymal stem cells; airway smooth muscle stem cells; vascular stem cells; plasticity; tissue repair and regeneration; lung stem cell related diseases; stem cell-based therapies.

4.1 Introduction

In the adult mammalian organs, there are some endogenous stem and progenitor cells which are distributed in the predetermined micro-environment named niche. Stem cells are a multipotent source of multiple cell lineages. They are undifferentiated cells that have the characteristic ability of self-renewal and development into various cell types carrying out different functions. Stem cells are an essential driving force for a number of rapidly growing scientific fields such as regenerative medicine and biology, and tissue engineering. In addition, stem cells play important functional roles during organ development, and in both tissue repair and regeneration after injury due to their pluripotency and ability to self-renew.

Researchers have reported that the endogenous pulmonary stem and progenitor cells exist in the respiratory tissues in adult humans and rodents. Lung-specific stem and progenitor cells play a critical role in the maintenance of pulmonary structural stability and functional repair (Yang *et al.*, 2016). Aging may be an important factor in the functions of stem and progenitor cells in lung repair since failure to repair and regenerate may inevitably occur with ageing. This may be due to the failure of endogenous stem and progenitor cell failures. Current research focused on lung endogenous stem and progenitor cells face many challenges, including a relative lack of specific molecular markers, difficulties of isolation and culture of these cells, and controversial views on their classification.

Remarkably, the lung is a good example of the organs that have a much slower turnover. This slow turnover has clearly hampered the labeling, identification and characterization of lung-specific stem and progenitor cells, unlike many examples of epithelial tissues undergoing rapid regeneration, such as the skin and gastrointestinal tract.

Thus, little is known about many lung-specific self-renewing stem and progenitor cells, in contrast to other well-studied organs. In addition, we do not fully understand whether a single lung stem cell suffices to produce more than 40 distinct cell types, which are required for the proper functions of the mature lung. Moreover, the lack of stem cell-specific markers and clonality assays for both the identification and isolation of these stem cells are among several other factors that also hampered progresses in the field of stem cell research in the lung.

Recent studies on the murine airway have shown that there are at least five putative epithelial stem and progenitor cell niches, beside both endothelial stem cells and airway smooth muscle (ASM) stem cells in the pulmonary vasculature. Another source of stem cells in the lung includes the circulating stem cells that probably take up residence in the lung (Hogan *et al.*, 2014; El-Hashash, 2018). In this chapter, we describe different types of endogenous stem and progenitor cells in the lung and the regulatory mechanisms of their behavior, including their proliferation and differentiation, and plasticity during pre-natal and post-natal development, morphogenesis, repair and regeneration of the lung.

4.2 Lung Stem and Progenitor Cell Types

4.2.1 *Embryonic epithelial and alveolar stem and progenitor cells*

4.2.1.1 *Localization and characterization*

Several studies have reported that distal tips of the branching epithelial lung tubules have many undifferentiated epithelial multi-potent stem and progenitor cells, which express several characteristic genetic markers at least at the pseudoglandular stage of lung development. Presumed endogenous epithelial stem/progenitor cells in the adult lung reside in the basal layer of upper airways within the alveolar epithelium, the bronchoalveolar junction, and within or near the pulmonary neuroendocrine cell rests, as shown in several studies (Engelhardt, 2001; Giangreco *et al.*, 2002, 2004; Reynolds *et al.*,

2000, 2004; Reddy *et al.*, 2004; Kim *et al.*, 2005; Rawlins *et al.*, 2006). It has been shown that cell cycle kinetics of distal epithelial cells differs from that of other epithelial cells, and most of these cells incorporate bromodeoxyuridine (BrdU), the analog of thymidine during one-hour pulse (Okubo *et al.*, 2005).

Furthermore, Rawlins and colleagues have shown that the distal lung epithelium has multi-potent progenitors, which contribute descendents to bronchi as well as alveoli during lung development (Rawlins *et al.*, 2009). Several lines of evidence have shown that descendants of distal lung epithelial progenitors are left behind in the stalks during epithelial branching, while proliferative stem cells stay in the epithelial budding tips. For instance, unique high expression patterns of the transcription factors sry box containing gene 9 (Sox 9), inhibitor of differentiation 2 (Id2), v-myc myelomatosis viral related oncogene (N-myc), and ETS variant gene 5 (Etv5/ERM) have been observed in the distal lung epithelial cells. Furthermore, distal lung epithelial cells have been regulated by several major signaling pathways, including, sonic hedgehog (Shh), Bone morphogenetic protein (Bmp), fibroblast growth factor (Fgf) and Notch as well as Wingless/int (Wnt) pathways (Shu *et al.*, 2005; Bellusci *et al.*, 1996; Liu *et al.*, 2002). Similarly, these signaling mechanisms are reiterated in other organs such as pancreas (Seymour *et al.*, 2007; Zhou *et al.*, 2007).

4.2.1.2 *Alveolar epithelial cell repair and regeneration*

It was proposed that decline in regeneration and repair of tissues during ageing could be a result of endogenous stem cell failure. The "ready-reserve" for replacement of injured alveolar surface has been proposed to contain large number of alveolar epithelial cells. For instance, acute oxygen damage of the alveolar epithelial cells leads to up-regulation of telomerase expression, which is a stem/progenitor cell marker (Driscoll *et al.*, 2000). It has been proposed that either alveolar epithelial cells contain a comparatively large sub-population of stem cells or most of alveolar epithelial cells have the capability to be reactivated into a stem cell-like state following tissue injury (Driscoll *et al.*, 2000).

Furthermore, bronchoalveolar stem cells (BASCs), which carry stem cell properties, have the capability to resist naphthalene-induced lung injury (Kim *et al.*, 2005). These BASC cells express epithelial cell markers, such as alveolar surfactant protein-C (SP-C) and airway Clara Cell 10 kD Protein (CC10), as well as co-express stem cell antigen-1 (Sca-1), and identified within or near bronchiolar–alveolar junctions. BASCs can undergo proliferation and differentiation into either alveolar cells or Clara cells. *In vitro* clonal assay of BASCs reveals self-renewal and differentiation as well as multi-potency. Additionally, several other studies have demonstrated that the variant Clara cells are endogenous lung stem cells which occasionally proliferate in steady-state conditions and may be responsible for repopulation of the distal airway epithelium as a response to tissue damage (Hong *et al.*, 2001). These variant Clara cells are identified by their capability to express Clara cell secretory protein. However, they differ from the more abundant Clara cells because they can survive against naphthalene-induced injury. Moreover, the kinetics of [3H] thymidine-labeled Clara cell in growing lung revealed that these cells are capable of self-renewal and act as progenitors for ciliated cells during the early postnatal period (McDowell *et al.*, 1985; Plopper *et al.*, 1992). Perl and colleagues have supported this finding by lineage labeling data (Perl *et al.*, 2005). However, further evidences are still needed. In addition, it has been elucidated that alveolar epithelial type II cells proliferate and result in type I cells following injury of adult alveoli. This is supposed to occur during postnatal growth (Evans *et al.*, 1975). The identification and characterization of various types of endogenous populations of alveolar stem and progenitor cells will lead to the identification of novel targets for rational regenerative lung therapies.

4.2.1.3 *Regulatory mechanisms of alveolar epithelial cells during lung development, repair and regeneration*

Some transcription factors have crucial roles in development, repair and regeneration of epithelial progenitor cell (Warburton *et al.*, 2010) (Figure 1; Table 1 in Chapter 1). The eyes absent homolog 1 (Eya1) and sine oculis homeobox 1 (Six1) are important transcription factors for the lung epithelial stem/progenitor cells maintenance. Mice deficient in

Eya1 or Six1 mice do not have the epithelial progenitor cell markers and highly express differentiation markers in their lungs. Additionally, their lungs are markedly hypoplastic with decreased epithelial cell branching capacity and intensive mesenchymal cellularity (El-Hashash *et al.*, 2011a, 2011b). Furthermore, E74-like transcription factor-3 (Elf3) has a pivotal function in the control of cell self-renewal and asymmetric cell division in the lung during recovery of the injured bronchiolar airway epithelial cells in response to specific injuries to Clara cells (Oliver *et al.*, 2011). Moreover, another research study has suggested that Thyroid transcription factor-1 (Ttf-1) can play a critical functional role in the modulation, and probably the initiation, of the early stage of compensatory growth in the lung (Takahashi *et al.*, 2010).

Other research studies have drawn attention to the functional roles of growth factors, including Fgf family members, in development of lung epithelial cells and protection of alveolar cells against lung damage (Ramasamy *et al.*, 2007). Interestingly, Fgf7 has been evaluated as a curative agent for animal models of alveolar injury (Ray *et al.*, 2003; Plantier *et al.*, 2007). Therefore, it has been hypothesized that preserving progenitor cell function using small molecules like inosine could be an effective therapeutic option in clinical trials. Buckley *et al.* have supported this hypothesis by showing that treatments with Fgf7 and inosine can lead to the alleviation of DNA damage in alveolar epithelial cells. These treatments can also enhance mitochondrial conservation and the alveolar epithelial cell migration and repair in a scratch assay in culture (Buckley *et al.*, 1997). In addition, treatment with inosine can potentially reduce the damage caused due to oxygen injury by glutathione repletion and decreased apoptosis and mitochondrial preservation as well as enhanced vascular endothelial growth factor (Vegf) expression (Buckley *et al.*, 2005). Additionally, Fgf10 has been found to have an important function in antagonizing lung injury and fibrosis (Gupte *et al.*, 2009). Other research studies showed that epithelial Fgf9 can mainly influences epithelial branching morphogenesis, while both mesothelial Fgf9 and mesenchymal Wnt2A are primarily responsible for maintaining signaling of mesenchymal Fgf-Wnt/ beta-catenin (Yin *et al.*, 2011).

4.2.2 Epithelial stem and progenitor cells of the bronchi and trachea

4.2.2.1 Localization and characterization

Studies on lung injury and repair models have identified multiple candidate endogenous stem/progenitor cells in the epithelium of trachea and bronchi. For example, Hong *et al.* have shown that subsets of keratin-14 (K-14)-positive basal cells (BCs) in the trachea can restore a differentiated epithelium following injury and these cells are different from BCs in the bronchi (Hong *et al.*, 2004). Furthermore, these K-14+ cells have the capability to function as progenitor cells; however, ciliated cells cannot, as demonstrated by several successful lineage-tracing experiments on the lung and trachea of adult mouse (Hong *et al.*, 2004; Rawlins *et al.*, 2007). In addition, Rawlins *et al.* have reported that Clara cells that express *Secretoglobin1a1* (Scgb1a1) can self-renew and differentiate because of tracheal injury. This ability for self-renewal and differentiation, however, is probably not the key mechanism of the regeneration of trachea (Rawlins *et al.*, 2009).

4.2.2.2 Repair and regeneration of epithelial stem and progenitor cells of the bronchi and trachea

A remarkable study from the Brigid Hogan laboratory has shown that specific epithelial stem and progenitor cells populations have a pivotal role in maintaining lung alveoli (Rawlins *et al.*, 2009). In addition, Rawlins *et al.* exploited the restricted expression of Scgb1a1, a marker of Clara cells, by producing a "knock in" transgenic mouse with a tamoxifen (TM) inducible Cre-recombinase (ScgB1a1-CreER™) resulting in lineage-labeled Clara cells of the airway. Using different doses and schedule of TM, these Clara cells were found to proliferate and differentiate into ciliated cells. Therefore, these Clara cells contribute to reconstitution of the epithelium during tracheal repair, while BASCs which co-express Scgb1a1 and surfactant protein C (SP-C) have no clear role in maintenance or regeneration of postnatal lung cells as previously proposed (Rawlins *et al.*, 2009). However,

Hong *et al.* have found that Keratin-14 (K-14)-positive BCs are alternative progenitor cells that contribute to self-renewal and proliferation of injured bronchial epithelium after depletion of Clara cells (Hong *et al.*, 2004).

Using lineage-tracing method, Hogan *et al.* have drawn attention to the behavior of a population of BCs as stem cells in the trachea. Considering that a population of BCs expressing cytokeratin 5 (Krt5) is present in the tracheal pseudostratified epithelium in mice and humans, a Krt5-CreER^{T2} transgenic murine line was used for lineage tracing experiments. This study suggested that BCs of mouse trachea can function as progenitor cells during postnatal lung growth and during the steady state (Rock *et al.*, 2009). They can also act as progenitor cells during the recovery of tracheal epithelial cells after sulfur dioxide-induced lung injury (Rock *et al.*, 2009). A clonality assay study from Rock's laboratory showed the importance of BCs in lung, as they are capable of self-renewal and differentiation into mucus and ciliated cells even in the absence of stromal or columnar epithelial cells in mouse and human airways. In addition, the transcription factor grainyhead-like 2 (GRHL2) was shown to play critical functions in the bronchial epithelial cell development, both as undifferentiated progenitor cells and as organized muco-ciliary epithelium (Gao *et al.*, 2013).

A diverse population of stem cells in lung can express both airway and mesenchymal origin molecular and genetic markers. These cells are characterized as a Hoechst dye-effluxing side population (SP) cells (Giangreco *et al.*, 2004). In addition, Hackett and colleagues showed the significant proliferative capacity of the CD45$^-$ SP cells, which are present in tracheobronchial epithelium of humans. They also suggest that many dysregulated pluripotent cells may have a principal functional role in the pathogenesis of major lung diseases such as asthma (Hackett *et al.*, 2008). Another study has demonstrated that some of both lung CD45$^+$ and CD45$^-$ SP cells have endothelial progenitor cell (EPC) characteristics following hyperoxic exposure during lung development (Irwin *et al.*, 2007).

Furthermore, cells from gland ducts are the main source of regenerated airway tracheal epithelium following induced injury (Borthwick *et al.*, 2001). Another study has shown that the submucosal gland (SMG) ducts found in the proximal airway probably contain putative stem cells (Lu *et al.*, 2002). Moreover, the functions of airway SMG duct cells in the repair of both the tubules of SMGs and surface epithelial cells (SECs) following sever hypoxia-induced ischemic injury has been uncovered (Hegab *et al.*, 2011). Using *in vivo* and *in vitro* stem cell model systems and lineage tracing, Hegab *et al.* (2011) showed that the cells of SMG duct have the capability to proliferate and differentiate into SMGs and SMG duct cells. They can also form the SE in the area adjacent to the sub-mucosal duct. They deduced that SMG duct cells are repairing stem cells for airway epithelium, which may play a pivotal role in the treatment of lung diseases as figuring out the repairing cell populations leads to discovering new therapeutic targets and novel cell-based therapies for airway diseases (Hegab *et al.*, 2011).

4.2.3 *Mesenchymal stem and progenitor cells*

Endogenous mesenchymal stem and progenitor cells are generally not well studied compared to epithelial stem and progenitor cells in the lung and other organs. Nonetheless, signals from lung mesenchymal cells were found to have a main role in branching morphogenesis and differentiation of lung epithelium. For example, beta-catenin signaling is essential for mesenchymal Fgf signaling, which in turn has a critical role in control of mesenchymal proliferation (Yin *et al.*, 2011). Moreover, the activation of Fgf10 signaling from the peripheral mesenchyme to epithelium is regulated by Fgf9 that is derived from the mesothelium. This regulation takes places through the assembly of signaling complex that consists of FGFR2b receptor, Grb2, SHP2, Ras and Sos in the epithelium as well as Sprouty-2, an inducible negative modulator of this critical signaling pathway for lung development (Bellusci *et al.*, 1997; Tefft *et al.*, 2005; Tefft *et al.*, 2002; del Moral *et al.*, 2006).

Glycogen synthase kinase-3beta/beta-catenin signaling has been considered as an important regulator of the differentiation of lung mesenchymal stromal cells (MSCs) of neonatal murine lung into myofibroblasts (Popova *et al.*, 2012). In addition, Six1/Eya1 signaling can control Shh signaling, which negatively regulates Fgf10, at normal level needed for appropriate lung growth. Thus, Shh expression increase above normal levels in the absence of either Six1 or Eya1, as well as its ectopic expression, inhibits Fgf10 signaling pathway, resulting in severe abnormalities of the lung mesenchyme and progenitors of epithelial branching. Moreover, T-box transcription factor 2 (Tbx2) is an important regulator of lung development, since its absence in murine lung was shown to lead to highly hypoplastic mesenchymal cells with severely decreased proliferation, but it prematurely induced mesenchymal differentiation into fibrocytes (Lüdtke *et al.*, 2013). Different types of endogenous mesenchymal stem and progenitor cells such as smooth muscle stem/progenitor cells and vascular stem/progenitor cells will be discussed in the following sections.

4.2.3.1 *Smooth muscle stem and progenitor cells*

In peripheral ASMs, Fgf10-expressing peripheral mesenchymal cells have been proposed to serve as progenitor cells for peripheral ASMs during early development of lung. Several studies showed that the ASM progenitors evolved as Fgf10-expressing cells, which distribute with the elongation of peripheral airway during the development of the airway, mimicking the action of a person pulling on a sock (Ramasamy *et al.*, 2007; De Langhe *et al.*, 2006; Mailleux *et al.*, 2005). These studies likewise have shown that Shh and Bmp4 signals are expressed in the distal airways and control the trans-differentiation of ASM progenitors and make them express alpha-smooth muscle actin fibers. In addition, a population of progenitors in ASM has been found to originate from the proximal mesenchyme (Shan *et al.*, 2008), and these are activated by Wnt signaling because Wnt2 signaling regulates expression of myocardin-related transcription factor-B (Mrtf-B) as well as Fgf10 in the lung (Goss *et al.*, 2011).

4.2.3.2 *Vascular stem and progenitor cells*

Some studies emphasize the importance of Vegf, erythropoietin and nitric oxide in on the mobilization and homing of lung EPCs during developmental changes such as bronchopulmonary dysplasia (BPD) (Stevens *et al.*, 2008). For example, Balasubramaniam *et al.* have found that decreasing the surface area of gas exchange in alveolar and vascular compartments occurs as a response to oxygen toxicity. These developmental disturbances have been associated with decreased expression of endothelial nitric oxide synthase, Vegf and erythropoietin receptor as well as reduced number of EPCs in both blood and bone marrow (Balasubramaniam *et al.*, 2007).

The differentiation of hemangioblasts to form a stereotypic capillary network that circles the bronchial, segmental as well as lobar branches of the airway was demonstrated by the activation of epithelial-derived Vegf (Ramasamy *et al.*, 2007; del Moral *et al.*, 2006). Appropriate organization of this vascular plexus may play an important function in correcting airway branching as well as tissue perfusion. Therefore, cross-talk between mesothelial, mesenchymal, epithelial and endothelial cells matches epithelial and vascular progenitor function and may play a critical role in lung repair and regeneration (Warburton *et al.*, 2010).

Interestingly, a population of progenitor cells was detected in the mesothelium overlying the lung that gives rise to pulmonary vascular smooth muscle cells during embryonic growth (Que *et al.*, 2008). However, endogenous circulating bone marrow progenitors or vascular wall progenitor cells may give rise to vascular EPCs. In addition, using a transgenic mouse line carrying a Bmp-responsive eGFP reporter allele showed that canonical Bmp pathway is active mainly within the ASM layer and the vascular network during the pseudoglandular stage of lung development (Sountoulidis *et al.*, 2012). However, further studies are still needed to find more lung progenitors in the other locations in the pulmonary vasculature.

4.3 Plasticity of Lung Stem Cells

The plasticity of adult stem cells is their abilities to have a remarkable mature phenotype, which is different from their tissue origin (Huttmann *et al.*, 2003). Many research studies have questioned the idea that various tissue types rely for their maintenance on their tissue-specific stem/progenitor cells only. Indeed, adult stem cells in different types of tissues and organs can generate both their own cell lineages and cell lineages of other tissues. In addition, these adult stem cells can sometimes pass through embryonic-derived barriers that were previously considered to be impermeable (Herzog *et al.*, 2003; Korbling and Estrov, 2003). For example, some controversial reports have suggested that adult stem cell types that do not belong to the bone marrow can reconstruct the hematopoietic tissues and *organs* that are involved in the production of the blood cellular components. In contrast, there are accumulated evidences that bone marrow-derived cells can generate various types of non-hematopoietic cells.

The lung has been considered as a primary site for the production of terminal platelets, with a remarkable haematopoietic potential, as shown in several recent lung studies (Lefrancais *et al.*, 2017; Tata and Rajagopal, 2017). Several human tissues such as the bone marrow, the adipose tissue, and the placenta are rich sources for the isolation of MSCs that can play an important role in the processes of lung injury, repair and regeneration (Rojas *et al.*, 2005; Lee *et al.*, 2009, 2011). For example, MSCs that are derived from the bone marrow can support hematopoietic stem cells (HSCs) and reside close to the sinusoids. In addition, they may play a protective role in bleomycin-induced lung injury in mice (Rojas *et al.*, 2005). Moreover, MSCs have the capacity to secrete different types of paracrine factors, which can control the permeability in both the epithelial and endothelial tissues. These factors can also function to inhibit both lung inflammation and bacterial growth, but increase tissue repair (Lee *et al.*, 2009, 2011).

There are controversial reports based on experimental studies that aim to understand the circulatory delivery of lung stem and progenitor cells. These reports are based on animals and human clinical studies and support the circulatory delivery of stem/progenitor cells in the lung, while others are against this delivery. It is well known that many bone marrow-derived cell types such as alveolar macrophages, mast cells and dendritic cells as well as lymphocytes, can normally migrate to the lung (Lefrancais *et al.*, 2017; Tata and Rajagopal,

2017). In contrast, other research studies have described that circulating cells can form many types of lung resident cells, including many epithelial, endothelial and myofibroblast cell types, under certain circumstances (Lefrancais *et al.*, 2017; Tata and Rajagopal 2017). Yet, there are several challenges facing the technical approaches used to identify these cells, including the co-localization of an inducible markers and clonal cells using lineage-tracing experimental approaches, and proteins that are characteristic of lung differentiated cell types such as collagen in fibroblasts or keratin in epithelial cell types.

The whole donor bone marrow that can grow in culture, MSCs and preparations enriched for HSCs are intensively used in many transplantation murine studies. Whole-body irradiation is commonly used for the depletion of the host bone marrow. However, it may also damage lung tissues. Remarkably, lung injury has been widely reported to apparently enhance the MSCs engraftment into lung (Theise *et al.*, 2002; Epperly *et al.*, 2003; Schmidt *et al.*, 2003; Davie *et al.*, 2004; Hashimoto *et al.*, 2004; Ishizawa *et al.*, 2004; Bentley *et al.*, 2010; Zhou and You, 2016). In addition, MSCs and whole bone marrow as well as HSCs can successfully reconstitute parenchymal cells in the lung (Grove *et al.*, 2002; Abe *et al.*, 2003; Ortiz *et al.*, 2003; Berger *et al.*, 2006; Chang *et al.*, 2009; Balduino *et al.*, 2012; Park, 2016).

The molecular and cellular mechanisms that are involved in MSCs transplantation and treatment in mice or rats have been intensively investigated using several *in vivo* models of lung diseases, repair and regeneration. For example, treatment with MSCs alone can significantly reduce acute pulmonary inflammation that is induced by LPS through the overexpression of angiopoietin 1 in mice (Mei *et al.*, 2007). In addition, treatments with type I and type II alveolar epithelial cells, MSCs that express fibroblast markers, myofibroblasts, have been successfully reported in bleomycin-induced lung injury mouse model (Rojas *et al.*, 2005). Moreover, the administration of MSCs may reduce the emphysematous changes that are associated with MSC differentiation into alveolar epithelial type II cells (Zhen *et al.*, 2008). Indeed, alveolar epithelial type II cells are epithelial stem and progenitor cells in the lung (Barkauskas *et al.*, 2013). Furthermore, the transplantation with HSCs were shown to yield up to 4% bronchial epithelial cells and 20% donor-derived pneumocytes in a murine animal model (Krause *et al.*, 2001). However, another study has identified only hematopoietic chimerism by HSCs (Wagers *et al.*, 2002), while whole bone marrow infusion can generate both fibroblasts and type I pneumocytes or type II pneumocytes (Grove *et al.*, 2002; Abe *et al.*, 2003).

Several studies have shown that radiation pneumonitis can augment whole bone marrow's generation of fibroblast cells (Epperly *et al.*, 2003), or both type II pneumocytes and bronchial epithelial cells (Theise *et al.*, 2002). Remarkably, bleomycin-induced lung injury can also stimulate the whole bone marrow to form more type I collagen-producing cells (Hashimoto *et al.*, 2004), while elastase-induced emphysema can promote both alveolar endothelial and epithelial cell formation (Ishizawa *et al.*, 2004). In addition, multiple studies have suggested that the lung-specific injury alone is sufficient to induce the migration of lung stem/progenitor cells (without bone marrow transplantation). Many circulating fibrocytes were indeed recruited into the bronchial tissues in the ovalbumin model of asthma (Schmidt *et al.*, 2003). In addition, many cells that can generate both endothelial and smooth muscle tissues have been detected in the circulation in a bovine culture model of hypoxic muscle cells *in vitro* (Davie *et al.*, 2004).

Both bone marrow transplantation and sex-mismatched lungs are excellent natural models for the analysis of cell behavior between the donor and recipient in humans. Interestingly, type II pneumocytes as well as both bronchial epithelial and gland cells of host origin were reported in one lung allograft research study (Kleeberger *et al.*, 2003). However, other studies could not detect these cell types (Bittmann *et al.*, 2001). In addition, the donor derived-epithelial cells could not be detected after a bone marrow transplantation in the nasal passages of the recipient (Khan *et al.*, 2010). Similar to lung allograft studies, both the endothelium and epithelium of donor origin have been detected in some studies (Suratt *et al.*, 2003), but not in other studies (Kleeberger *et al.*, 2003), following the transplantation of bone marrow.

A number of studies that focus on the fusion of MSCs and lung epithelial cells have shown that cell fusion takes place *in vivo* and in culture, which may explain the existence of both donor and lung cell markers in some cells (Spees *et al.*, 2003; Camargo *et al.*, 2004; Wagers and Weissman, 2004). Alternatively, the cells may be reprogrammed in the lung environment during their transdifferentiation, which is characteristic to cells that belong to a specific tissue type and are able to differentiate to a cell type of another tissue. Interestingly, many events that are attributed to trans-differentiation probably represent the cell fusion process (Camargo *et al.*, 2004).

Many pluripotent stem cell (PSC) types have been discovered in the bone marrow (Jiang *et al.*, 2002). Yet, littles known about the bone marrow cell components that are crucial for the engraftment of lung. In addition, it is possible that rare trans-differentiation, which represents migrating bone marrow pluripotent cell types that resemble embryonic germ cells or embryonic stem cells (ESCs), still exist in the adult bone marrow. Moreover, it is still not clear whether cells from the bone marrow must undergo transition via intermediate compartment(s) before their colonization of the lung. It, therefore, is still not known whether cells of the bone marrow should pass through a middle compartment(s) before colonization of the lung, or the circulating stem cells move from other sources rather than the bone marrow. Remarkably, typical hematopoietic cell lineages that are derived from the bone marrow as well as chimeric cells resulting from fusion or by lung cells generated by trans-differentiation may have the ability to promote the local formation of more stem cells or lung-specific repair function during the processes of lung repair and regeneration (Yoder and Ingram, 2009).

Some research studies have suggested that MSCs can reduce both collagen concentration and Smad2 phosphorylation, suggesting that these cells have certain anti-fibrotic properties (Rojas *et al.*, 2005), while cells of the bone marrow can effectively contribute to the fibrotic tissue formation (Hashimoto *et al.*, 2004). In addition, the bone marrow-derived cell types probably play mitigating or exacerbating roles in the processes of lung repair or fibrosis (Sage *et al.*, 2008). Whether the ability of repairing lung injuries is dependent on circulating cells or whether the exogenous delivery of cells can increase the efficiency of resisting lung injuries or promoting the healing process are still controversial research topics that need more studies.

4.4 Derivation of Lung Epithelial Cells in Culture

In vitro induced human and murine ESCs can express several markers such as surfactant proteins as well as form both lamellar bodies and pseudoglandular structures, which imply alveolar epithelial type 2 (AT2) cell phenotype (Kotton *et al.*, 2005; Wang *et al.*, 2007; McIntyre *et al.*, 2014). Remarkably, ESCs grown *in vitro* under air–liquid interface conditions can exhibit many specific genetic markers of the airway alveolar epithelium (Van Haute *et al.*, 2009; Sadeghian *et al.*, 2016). Recent lung research has shown that both human embryonic lung epithelial cells and murine embryonic lung epithelial cells are multi-potent progenitor cells that can be successfully expanded *in vitro* (Nikolić *et al.*, 2017). However, this recent research is still limited since they focus on using a limited number (one or two) of immune-phenotypic markers, including surfactant proteins, and the derived airway or alveolar cells have not been shown to function properly yet. Interestingly, more successes have been achieved using recent and more effective research protocols combining more complex understanding and applications of cellular signaling pathways to guide the development of embryonic lung.

Furthermore, some lineage-tracing approaches and tools that have been recently developed, such as Nkx2.1-GFP-expressing mouse models, have produced more robust data on the cell derivation with the airway phenotypes that characterize both type 1 (AT1) and type 2 (AT2) alveolar epithelial cells from human/mouse ESCs and from inducting pluripotent stem cells (iPSCs) in culture (Roszell *et al.*, 2009; Green *et al.*, 2011; Mou *et al.*, 2012; Longmire *et al.*, 2012; Wong *et al.*, 2012). Since these derived cells are capable of repopulating the scaffolds of decellularized whole lungs, they can therefore be used for the regeneration of whole lung (Longmire *et al.*, 2012). In addition, recent advances in both iPSCs and ESCs and their applications can facilitate our understanding of molecular mechanisms controlling the processes of lung injury and repair. However, further research studies are still needed to support the applications of both iPSCs and ESCs in the clinical treatments of different types of lung diseases.

4.5 Lung Stem Cell Diseases

4.5.1 *Stem cell-related diseases in the lung*

Some well-known lung diseases probably involve lung-specific stem cells. These diseases can be simply and roughly classified based on the defects of their stem cell behavior, such as the abnormal increase or decrease of stem cell proliferation/self-renewal. For instance, the pathophysiology of adult respiratory distress syndrome (RDS) is probably related to an impairment of the barrier function of pulmonary epithelial and/or endothelial cells.

Remarkably, one approach that probably improves the barrier functions of epithelial and endothelial tissues is the transfer of endogenous epithelial/endothelial stem cells or the transfer of other types of stem cells, including ESCs, adult somatic stem cells, or embryonic germ cells, which also supports the concept of "stem cell deficiency" (Easleyet *et al.*, 2014). In addition, a destruction of the bronchiolar epithelium, whether it is viral, toxic or allo-immune, suggests a deficiency of stem cells in bronchiolitis obliterans (Zemke *et al.*, 2009). In contrast, other lung diseases are generally considered to be due to a hyper-proliferation of fibroblast stem cells and include both fibrotic reactions and scarring in response to epithelial injury (Xu *et al.*, 2008). Stem cell augmentation has been generally thought as the potential mechanism for minimizing lung injury, augmenting lung repair and regenerating injured lung tissues. The other view is that that the suppression of excessive stem cell growth is probably a potential effective therapeutic approach when excessive lung cell proliferation leads to the pathophysiology of some lung diseases, such lung cancer, fibrosis or smooth muscle hyperplasia, and therefore the inhibition of stem cell overgrowth may be an effective therapeutic target.

4.5.2 *Mesenchymal/Stromal stem cells and the immunomodulation of lung diseases*

Pluripotent ESCs may be used as infinite resources for therapeutic applications of different types of human diseases (Pera and Trounson, 2004). Adult somatic stem cells play important roles in the maintenance of cell populations in different adult tissue types (Schira *et al.*, 2015; Tweedell *et al.*, 2017). Mesenchymal stem cells (MSCs) are a good example of adult or somatic stem cells that are important in the immunomodulation of different lung diseases.

The control of different sources of MSC such as the bone marrow, fat and placenta during both inflammatory and immune diseases has been well investigated (Keating, 2012; Borger *et al.*, 2017). Similar to adult MSCs, ESCs-MSCs or iPSCs-MSCs also have a great potential for immunomodulation by inhibition of cytokine profiles, micro-environment exosomal modulation and secretion of bioactive paracrine factors (Tan *et al.*, 2015; Sabapathy and Kumar, 2016). Besides the paracrine effects of many soluble peptides or other mediators, recent data indicate that the release of MSC-attached or microsomal granules can affect the behavior of the nearby structural and inflammatory cells (Gao *et al.*, 2016). In addition, MSCs can effectively serve as antigen-presenting cells. They can also transfer both mitochondria and other cytosolic components via connexin bridges (Prockop and Oh, 2012). Moreover, MSCs may be capable of promoting lung repair by the stimulation of endogenous lung-specific stem/progenitor cells such as bronchioalveolar stem/progenitor cells in a murine model of BPD (Tropea *et al.*, 2012). Notably, MSCs are widely considered as a vector for delivering of many therapeutic genes and proteins since they can be transduced through multiple transfection or transduction pathways. Interestingly, many clinical trials on autoimmune and inflammatory diseases have shown that allogeneic MSC administration appears to be safe and feasible in other tissue types (Colmenero and Sancho-Bru, 2017). In addition, MSCs are effective potential therapeutic for inflammatory diseases in other organs such as in an inflammatory bowel (Adak *et al.*, 2017).

Almost a 100 publications so far have demonstrated the successful systemic or intra-tracheal administration of MSCs in rodents and other pre-clinical animal models of different types of lung diseases, such as bronchiolitis obliterans, pulmonary hypertension, acute lung injury, BPD, chronic obstructive pulmonary disease (COPD), bacterial lung infection, fibrosis, asthma and obstructive sleep apnea (Li and Wu, 2015; Moodley *et al.*, 2016). Moreover, the administration of MSCs that are derived from bone marrow or placenta has successfully reduced both inflammation and injury in bacterially or endotoxin-injured lung explants in humans (Lee *et al.*, 2013). In addition, MSCs may contribute to the improvement of both pathological and functional outcomes in pulmonary fibrosis (Yao *et al.*, 2013; Liu *et al.*, 2016). Similarly, human MSCs can be used effectively in models of lung injury that include both immunodeficient and immunocompromised mice (Kim *et al.*, 2011). However, more research studies are still needed to determine the molecular mechanisms of MSCs as well as to optimize dosing regimens of stem cells for different clinical applications in the future.

Much progress have been achieved recently in understanding the roles of MSCs in different lung defects and disorders. However, the molecular and cellular mechanisms that regulate the roles of MSCs in different lung defects/disorders are not completely understood. These mechanisms probably vary due to the different immune and inflammatory environments that characterize each lung disease or disorder (Weiss, 2014). In addition, several studies have shown that after systemic administration of MSC, these stem cells are initially located in the lungs. In addition, lung injury could enhance localization and/or retention of MSCs that are derived from the bone marrow in the lung, which may eventually trigger the cells to have functional effects such as the production of anti-inflammatory proteins such as TSG-6 (Lee *et al.*, 2009; Weiss, 2014). Other studies have suggested that administrating conditioned media that are obtained from MSCs could mimic many of the ameliorating effects, which can result from MSC administrations in different types of lung injury models (Ionescu *et al.*, 2012).

A potential molecular mechanism for the effect of MSCs on lung inflammation and injury is that these stem cells have the potential of effective interactions with the immune system rather than direct actions on the lung. This potential molecular mechanism is supported by some studies which have shown that adult MSCs can effectively influence the response of immune T and B cells. For instance, adult MSCs could effectively inhibit both the proliferation of T-cells and production and cytotoxicity of cytokines. In addition, adult MSCs can control the balance of Th1/Th2 by promoting a Th1 phenotype *in vivo* (Glennie *et al.*, 2005; Puissant *et al.*, 2005) and effectively regulate the functions of Tregs by enhancing T-regulatory cells (Selmani *et al.*, 2008). Moreover, MSCs have a greatly inhibitory effect on the processes of dendritic cell maturation and activation as well as antigen presentation (Burchell *et al.*, 2010). Indeed, MSCs were shown to stimulate B cell viability, but can also inhibit their proliferation and block the cell cycle (Corcione *et al.*, 2006). Furthermore, MSCs can effectively change both the antibody production and B cell costimulatory molecules (Corcione *et al.*, 2006). Interestingly, adult MSCs have a suppression effect on the activity of interleukin-2 (IL-2)-induced natural killer (NK) cells (Spaggiari *et al.*, 2006). MSCs are, therefore, able to produce different effects that have important clinical applications in various types of lung injuries, diseases or disorders.

Other cell types can be used in the cell-based immunomodulation of lung diseases or disorders. For example, populations of human amniotic fluid cells, mononuclear cells derived from bone marrow or human amnion epithelial cells can effectively reduce lung injury in immunocompetent murine models (Araujo *et al.*, 2010; Hodges *et al.*, 2012).

4.5.3 *Stem cell-based therapies of different lung diseases*

Rapid progresses have been recently achieved in the field of cell-based therapy of different lung diseases or disorders, and these are both for lung repair and regenerative in nature. For example, many pre-clinical studies support the application of EPCs in different types of lung disorders or disease such as pulmonary hypertension. Moreover, an increased number of pre-clinical literature supports MSC applications in acute lung injury, inflammatory critical diseases and immunomediated conditions such as asthma and BPD as well as bronchiolitis obliterans. Many pre-clinical models of lung diseases could not completely mimic the pathogeneses of human lung diseases or predict the associated clinical changes (Matute-Bello *et al.*, 2011). This leads to a slow progress of clinical research that aims to develop cell-based therapies for different types of lung diseases or disorders. However, some recent clinical trials have provided strong evidence of the safe applications of MSCs in different types of lung diseases or disorders, such as multiple infusions in patients with chronic lung diseases. For example, a multi-center, placebo-controlled Phase II trial of systemic administration of MSC preparation, which is derived from bone marrow, has confirmed safety with no acute infusional toxicity in patients with chronic obstructive pulmonary disease (COPD) (Weiss *et al.*, 2013).

Remarkably, MSC administration is not probably the best therapeutic target at present for some lung diseases such as chronic persistent lung disorders or diseases with a low inflammatory level such as COPD. This also includes other diseases, including idiopathic pulmonary fibrosis (IPF), in which the pre-clinical data have shown that that MSCs are not effective (Toonkel *et al.*, 2013; Weiss and Ortiz, 2013). In contrast, other lung disorders or diseases are better targets for MSC interventions such as severe asthma, ARDS and sepsis/septic shock (Matthay *et al.*, 2010). A number of clinical trials of MSCs for ARDS or septic shock are currently underway in different institutes worldwide. The Clinical Trials.gov website shows a list of these MSC clinical trials, in addition to other clinical trials using EPCs in many lung diseases or disorders.

4.6 Stem Cells in Lung Repair and Regeneration

Many factors can cause lung injuries either directly or indirectly. For instance, as the lung exchanges gas directly with ambient environment, wounds caused by potential risky agents than can damage the lung, such as oxidant chemicals and enzymes performing proteolysis, have severe effects on the lung. In addition, the causal oxide and structural components of protease wear and tear like elastin and collagen can adversely decline age-related pulmonary function in normal individuals (Mahler *et al.*, 1986, 2003). Other processes and elements that might cause lung injury and/or damage include several chronic and acute lung diseases or their medical treatments via both oxygen and proper positive-pressure ventilations, which may lead to pulmonary damages more than the orderly repair capacities. This can lead to several major pathological changes such as the severe damage of tissues or fibrotic scarring (Jeffery, 2001; Ozkan *et al.*, 2001; Chow *et al.*, 2003).

Despite all these damaging processes and factors, lungs are capable of functioning well in normal individuals, indicating that lungs have their own repair capacity. The capacity of the lungs for repair or regeneration is determined by key factors. These key factors include the ability of lung-specific stem cells to proliferate and differentiate, and/or the re-entry of the remaining cells to the cell cycle to replace damaged cells or tissues (Warburton *et al.*, 2010; Zhu *et al.*, 2018).

The common thought is that tissues that undergo a self-renewal are apparently imbued with adult somatic stem/progenitor cells that are resident and tissue-specific stem cells. However, some studies show controversial evidences which indicate that stem/progenitor cells from one tissue type are probably able to form different cells types of other organs (Nirmalanandhan and Sittampalam, 2009; Little, 2011; Lefrancais *et al.*, 2017). For example, circulating cells that are derived from the bone marrow probably increase the resident stem/progenitor cells in other organs (Little, 2011). Moreover, recent studies have shown that the well-organized lung acts also as a source of hematopoietic stem/progenitor cells and a site for biogenesis of the platelets, suggesting that the lung may have functions similar to the bone narrow (Lefrancais *et al.*, 2017).

4.7 Regenerative Medicine in the Lung: Progress and Challenges

Recent data show that pulmonary disorders and diseases are the most important factors that can cause human morbidity and mortality followed by both cancer and cardiovascular diseases. Many research studies have successfully used growth factors for directing ESC differentiation and for inducing iPSCs into lung-specific epithelial cells that could be eventually used for the treatment of lung injury. For example, endodermal epithelial cells could be derived from PSCs by using activin to induce the inhibition of the activities of both BMP signaling and transforming factor-β pathways (TGF-β; Green *et al.*, 2011). In addition, the infusion of stromal cell from different human tissue types such as human bone marrow, cord blood, adipose tissues or placental tissues, which are referred to either as MSCs or as marrow stromal cells, are currently under way or being planned for the treatment of patients with BPD, COPD, acute lung injury, asthma or bronchiolitis obliterans. Notably, the infusion of endothelial stem or progenitor cells apparently have some paracrine and/or angiogenic effects on the receptor tissues (Weiss *et al.*, 2011). In contrast, the marrow-derived endothelial precursor cells may not contribute to the repair or maintenance of the lung endothelial tissues (Ohle *et al.*, 2012). Further studies are still needed to determine whether these infused cells can form new endothelial cells in the lung.

Stem cell therapy has many prospects in the lung. For instance, intensive research is still needed to both identify and characterize different types of lung disorders or diseases, in which the tissue damage takes place and exceeds the capacity for timely endogenous repair. In addition, more advanced studies are required to establish a standard source of stem or progenitor cells and identify new methods for delivering them to the appropriate compartment in the lung. Moreover, the potent self-renewal, proliferation and differentiation potentials of various types of stem cells such as ESCs, MSCs or iPSCs should be fully characterized and understood to enable good applications of stem cell therapy in the lung.

More research is still also needed to compare different preparations of MSCs that are used in systemic or intra-tracheal administrations directly. More studies are also needed to explore the differences between syngeneic, allogeneic and xenogeneic administration of MSCs in pre-clinical lung injury models. In addition, further studies on MSCs are also required to define optimal cell preparation and storage conditions as well as dosing, vehicle buffers and administration routes, i.e. direct airway vs. systemic.

There are several current challenges for stem cell therapy of major lung disorders or diseases. For example, it is widely known that the delivered therapeutic cells that are used in the lung injury or other tissue injury return to the desired microscopic site before their integration into the injured tissue to exert beneficial functions (Takahashi *et al.*, 2003). There are, however, potentials for some adverse effects of this cell therapy approach. In addition, the regeneration of alveolar septa is probably required for the repair of chronic lung diseases in premature infants and in emphysema (in adults) as well as for reversing lung dysplasia and hypoplasia. However, the regeneration of alveolar septa is currently difficult to be developed and achieved.

Furthermore, heterologous stem cells, or gene-corrected autologous stem cells are also required for different processes that are involved in pre-natal and post-natal development in the lung. For example, for an effective stem cell-based therapy of cystic fibrosis, heterologous or autologous stem cells are needed for airway colonization, as well as for both airway cell proliferation and differentiation into columnar cells that cover the airway lumen. Interestingly, reports suggest that some circulatory cells have the potential of generating single and isolated airway BCs that have long-term self-renewal and differentiation abilities (Rock *et al.*, 2009). In addition, stem cells with potentials for restoring the functions of alveolar endothelial and epithelial cells in the injured lung are essential for developing an effective stem cell therapy of the RDS.

Despite the common thought that lung injury promotes the recruitment of stem cells, more research is still needed to determine whether the process of stem cell recruitment is quick enough for effectively reversing acute and other cellular dysfunctions in the lung such as in lungs with RDS. In addition, stem cell drug therapies that are designed to modulate the regulatory signaling pathways of the development different types of lung diseases are the future direct goals and targets to promote and enhance lung repair and regeneration in humans. More accelerated efforts could be increased to bring stem cell and regenerative therapies for a wide range of lung diseases and disorders into clinical trials in different ways such as by enhancing lung research funds from different agencies and the public since progresses in basic science laboratories will eventually lead to more advances in clinical medicine related to lung diseases and disorders.

4.8 Summary and Conclusions

The lung is a complex organ with three distinct regions, trachea, bronchioles and alveoli, and over 40 different types of cells, including stem, progenitor and differentiated cells, that facilitate both the exchange and transport of gases. This complexity and the low levels of cell turnover has hindered the study of lung stem cell behavior, biology and function. Accumulated knowledge on potential lung-specific stem/progenitor cells in the lung has clearly enhanced in the recent years. Recent lineage-tracing studies have led to the identification of many putative populations of stem and progenitor cells in the lung that are involved in tissue-specific homeostasis and repair. These populations of stem cells and their differentiated progeny can display a notable lineage plasticity after lung injury. Thus, many endogenous epithelial cell types in the lung can function as differentiated cells that perform specific functions and act as transit-amplifying progenitor cells with a proliferation capacity in response to a specific lung injury.

Stem/progenitor cell populations in the lung include epithelial and alveolar stem/progenitor cells, stem and progenitor cells of the bronchi and trachea, and lung mesenchymal stem and progenitor cells such as ASM stem and progenitor cells and vascular stem and progenitor cells. Although they are of major interest, both the localization and properties of lung-specific stem cell niches and cell types within each niche are still a controversial topic. Identifying new stem cell populations and characterizing resident stem and progenitor cells in the lung are two essential steps to understand the repair/regeneration of lung after injury and lung diseases.

Lung-specific diseases are one of the major causes of mortality and morbidity in human infants and adults. Lung stem cell research and regenerative medicine are fast growing and can provide effective solutions to major lung disorders and diseases. The *in vivo* cell proliferation rate is generally low in the lung, compared to other systems. However, this rate is enhanced after lung injury, which can activate stem and progenitor cell populations and promote their division. In addition, stem cells are likely involved in some major lung diseases, and lung stem cells may have a major role in the repair and regeneration of the lung and in lung diseases. Further studies are still needed to explore stem cell-based therapies in lung disorders and diseases.

Chapter 5

Pattern Formation of the Anterior Endoderm and Developing Lung

Abstract

The germ layer of the endoderm can contribute to the formation of both the gastrointestinal and respiratory tracts, and other associated organs. The endoderm is generally responsible for the formation of the internal epithelial tube that will eventually become the digestive tract. During embryogenesis, the endoderm represents the inner germ layer in both triploblastic and diploblastic embryos. The anterior–posterior (A–P) and proximal–distal (P–D) patterning are among the earliest developmental events during embryogenesis. They are tightly regulated with a highly coordinated network of several signaling molecules and pathways. Accumulated data in the last two decades from studies on animal model organisms have enhanced our understanding of the anterior endoderm development and patterning and P–D patterning of the lung. These data have also uncovered many of the molecular mechanisms and signaling molecules that regulate these processes. In this chapter, we will describe this progress with a focus on the anterior endodermal patterning and its regulatory molecular mechanisms and signaling pathways, as well as the P–D patterning of lung embryonic cells. Lastly, we discuss the role of stem and progenitor cells in the P–D patterning of the lung.

Keywords: Endoderm; lung; patterning; signaling pathways; proximal-distal patterning; stem and progenitor cells.

5.1 Development and Patterning of the Anterior Endoderm

During embryogenesis, the endoderm represents the inner germ layer in both triploblastic and diploblastic embryos. The endoderm is generally responsible for the formation of the internal epithelial tube that will eventually become the digestive tract. The germ layer of the endoderm can, therefore, contribute to the formation of both the gastrointestinal and respiratory tracts, and other associated organs (Zorn and Wells, 2009; Hashimshony *et al.*, 2015; Figure 1 in Chapter 1).

Several gene-expression analysis studies in different organisms suggest the endoderm as an original germ layer (Hashimshony *et al.*, 2015). The lateral extensions of the developing endodermal tube in vertebrates normally form several organs such as the thymus, thyroid gland, lungs and pancreas as well as liver, while the outer ectodermal germ layer of the embryo normally gives rise to the neural crest cells, nervous system and skin (Zorn and Wells, 2009; Hashimshony *et al.*, 2015; Figure 1 in Chapter 1). Many other organs are derived from the third germ layer, the mesoderm, including the muscle, blood, cardiovascular system and kidneys. Evidences from molecular biology and fate mapping studies suggest the mesendoderm as the common progenitor for both the mesoderm and endoderm (Zorn and Wells, 2009).

Many major transcription factors for both the endoderm and mesoderm can be detected in the same cells in different organisms, including *Xenopus* and zebrafish (Lemaire *et al.*, 1998; Warga and Nusslein-Volhard, 1999; Rodaway *et al.*, 1999). In addition, fate mapping studies on the embryos of *Caenorhabiditis* show evidences that one blastomere forms both the endodermal and mesodermal lineages (Maduro *et al.*, 2001).

The differentiation of endoderm is dependent on the integration of many signaling molecules and events during early embryonic development (Table 1 in Chapter 1). For example, members of the transforming growth factor β (TGFβ) family of growth factors and signaling molecules such as Nodal is a major regulator of the mesendodermal development (Shen, 2007). In addition, nodal signaling controls both the endodermal cell segregation from the mesoderm, and the extra-embryonic endodermal patterning (Chen and Schier, 2001; Kruithof-de Julio *et al.*, 2011; Liu *et al.*, 2012). In addition, nodal signaling regulates the expression levels of several transcription factors that are critical for the endodermal cell differentiation, including Sox, Forkhead, PouV, Mixer and Gata transcription factors (Zorn and Wells, 2009, 2007; Chiu *et al.*, 2014). Nodal expression is dependent on retinoic acid (RA) activities in the mouse, and this RA-nodal signaling pathway functions to prevent the duplication of the body axis (Uehara *et al.*, 2009; Kelly and Drysdale, 2015). The role of this signaling pathway in mesendodermal cell differentiation is still not clear (Uehara *et al.*, 2009). In addition, the maintenance of nodal expression levels is dependent on Wingless/int (Wnt) signaling that shows a reciprocal interaction with nodal, which is essential for the differentiation of the endoderm (Liu *et al.*, 1999; Brennan *et al.*, 2001; Kelly and Drysdale, 2015).

The anterior–posterior (A–P) and proximal–distal (P–D) patterning are among the earliest developmental events during embryogenesis (an example for the lung is shown in Figure 1 in Chapter 1). They are tightly regulated with a highly coordinated network of several signaling molecules and pathways. A combination of several molecules and signaling pathways, including Bone morphogenetic protein (BMP), fibroblast growth factor (FGF) and Wnt, at the endoderm posterior end initiates the early steps of the endodermal A–P patterning (Zorn and Wells, 2009). The antagonizing of Wnt signaling at the endoderm anterior end can counteract these signaling molecules/pathways, leading to a proper transcription factor expression pattern that is required for normal endoderm formation and patterning (Zorn and Wells, 2009; Kelly and Drysdale, 2015). Thus, the expression of transcriptions factors such as Sox2 and Hhex marks the presumptive foregut and Pdx1 expression marks the midgut, while the expression of Cdx transcription factor marks the hindgut (Zorn and Wells, 2009). These signaling molecules and molecular mechanisms will be described in detail in the following sections and are summarized in (Table 1 in Chapter 1).

5.2 Molecules and Signals Regulating the Patterning of the Anterior Endoderm

Many mesodermal-derived signals and molecules produced near the anterior foregut play a critical role in both the specification and differentiation of endodermal stem/progenitor cells in the lung (Figure 2; Table 1 in Chapter 1). The developing anterior foregut can generate many tissue types such as the trachea, liver, lungs, esophagus, thyroid, biliary system, stomach and pancreas. One of the most interesting questions to both stem cell and developmental biologists, which still needs more research, is how mesoderm-derived signal pathways can promote distinct organ-specific stem/progenitor cells along the A–P axis. Many well-known signaling pathways have been implicated or determined in the regulation of early specification of foregut organs through distinct mesodermal–endodermal signaling interactions such as Notch, Wnt, FGF, sonic hedgehog (Shh), BMP, and RA (Zorn and Wells, 2009).

Nodal is a well-characterized member of the TGF-β signaling network, and among the earliest studied signaling molecules that are important for the early development of a definitive endoderm. In addition, the nodal signaling pathway is a critical signal from the developing primitive streak, which is essential for the induction of murine endoderm formation (Zorn and Wells, 2009). Shortly after the definitive endoderm formation and subsequent foregut tube formation, the A–P patterning of the endoderm occurs by signals produced by the surrounding mesenchyme such as Wnt, RA, FGF, Shh and BMP signaling pathways (Zorn and Wells, 2007, 2009). These signal pathways can trigger the expression of several transcription factors in the posterior endoderm, including Sox2, FoxA2 and Hhex that can define the developing anterior endoderm, and the expression of Cdx (Cdx1/Cdx2/Cdx4) family of transcription factors in the newly formed posterior endoderm (Zorn and Wells, 2007, 2009). Notably, the expression of these signaling pathways and factors often occurs in gradients in the developing endoderm. This gradient expression was reported to be sufficient to promote organ-specific endoderm progenitor cells of the foregut through a dose-dependent response along the A–P axis (Zorn and Wells, 2007, 2009).

During early embryogenesis, a high level of Wnt signaling activity, together with active FGF4 signaling, can induce the expression of Cdx, a key posterior endoderm marker, and suppress the genetic markers of the anterior endoderm fates (FoxA2 and Hhex) at E7.5–8.5 days of murine development (Zorn and Wells, 2009; Sherwood *et al.*, 2011). After the formation of the endodermal-derived foregut and hindgut, several localized signaling factors function to further refine the anterior endoderm to express specific transcription factors such as Pdx1 (pancreas), Nkx2.1 (lung) and Hnf4 (liver), which can promote and mark organ-specific domains (Zorn and Wells, 2007, 2009). In addition, a remarkable gradient of active FGF signaling from the neighboring cardiac mesodermal cells functions to promote further foregut patterning after E8.5 in the mouse embryo. In addition, increased FGF2 expression stimulates Nkx2.1 expression in the stem/progenitor cells of both the thyroid and lung, while lower FGF signaling levels can upregulate the expression of liver-specific genes (Zorn and Wells, 2007, 2009). The RA signaling pathway is a major regulator of the formation of several endodermal-derived tissues and organs, including the lungs, dorsal pancreas and stomach, probably by repressing the activity of TGF-β signaling pathway (Desai *et al.*, 2006; Kelly and Drysdale, 2015). The balance between these different signaling pathways leads to the proper specification of lung endodermal stem/progenitor cells within a specific region of the anterior ventral foregut between stem/progenitor cells that acquire the fate for thyroid and liver organs.

The foregut patterning along the dorsal–ventral (D–V) axis is also tightly controlled by signaling pathways from the neighboring mesenchymal cells. The D–V patterning of foregut endodermal stem/progenitor cells is also tightly regulated by signaling molecules from the surrounding mesenchymal cells like the A–P patterning. This tight regulation starts early during embryogenesis. For example, Wnt/β-catenin signaling pathway functions to suppress anterior endodermal cell fate and induce posterior endoderm identity at very early stages of murine foregut development (E7.5–8.5), and then effectively regulates the patterning of the anterior endoderm at E9–9.5 of mouse embryogenesis (Sherwood *et al.*, 2011). In addition, Wnt2/2b and beta-catenin signaling activates are critical in the regulation of the specification of lung stem/progenitor cells in the foregut (Goss *et al.*, 2009). Mesenchymal Wnt2/2a signaling stimulates Nkx2.1 expression in the ventral wall of the endoderm of the anterior foregut (Goss *et al.*, 2009). The loss of these two ligands eventually leads to a respiratory agenesis without changing the specification of progenitor cells of other organs in the muse embryo (Goss *et al.*, 2009). Similarly, β-catenin loss in the endoderm of the anterior foregut early during embryogenesis can also lead to a complete respiratory agenesis at E8.5–9.0 of development. This indicates the importance of canonical Wnt/β-catenin-signaling pathway in lung stem/progenitor cell specification in the developing anterior foregut (Harris-Johnson *et al.*, 2009; Goss *et al.*, 2009). This conclusion is supported by the ectopic generation of Nkx2.1+ lung endodermal stem/progenitor cells when a constitutively active β-catenin protein is expressed in posterior foregut regions (Harris-Johnson *et al.*, 2009; Goss *et al.*, 2009; Zorn and Wells, 2007, 2009).

The BMP is another signaling pathway that has spatial-specific and temporal functional roles in regulating both the specification and differentiation of foregut endodermal cells. BMP signaling activities from the ventral mesenchyme to the developing endoderm are a critical regulator of the D–V patterning of the anterior foregut during early murine embryogenesis (E8.75–E9.5 of development). In addition, BMP signaling pathway functions through specific receptors, such as BMPR1A/B, to inhibit the ventral expression of Sox2 that leads to the promotion of trachea/respiratory cell lineage formation (Domyan *et al.*, 2011). This BMP-regulated D–V patterning of the anterior endoderm can result in the inhibition of Sox2 gene expression in the ventral regions of the anterior foregut, which allows the stimulation Nkx2.1 expression and subsequent progress in the lung development (Domyan *et al.*, 2011). The proper specification of lung endodermal cells , therefore, requires a tight cooperation between both Wnt and BMP signaling pathways since the abilities of Wnt signaling to promote ectopic Nkx2.1+ lung endodermal stem/progenitor cells in the posterior foregut is largely dependent on and requires active BMP signaling at these early stages of embryogenesis (Domyan *et al.*, 2011). Remarkably, active BMP signaling functions to inhibit the abnormal ectopic lung budding from more posterior regions of the developing foregut, indicating the critical functional roles of active BMP signaling in the regulation of both lung identity and lung endodermal cell fate in specific regions that are dedicated to the formation other organs derived from the foregut (Domyan *et al.*, 2011). The changing functional role of two important signaling pathways for the patterning of the foregut endoderm, BMP and Wnt/β-catenin pathways, strongly indicates the importance of both mesenchymal–endodermal communications and the precise timing of active signals from the neighboring mesenchymal cells to the developing endodermal cells during early embryonic development.

In addition, Fstl1 gene is a BMP 4 signaling antagonist. Fstl1 deletion results in several defects in the respiratory system, as evident by malformed tracheal rings that are manifested as reduced ring number and discontinued rings as well as defects in the maturation of the alveoli (Geng *et al.*, 2011).

The formation and development of lung buds are also regulated by FGF10 signaling activities from the developing splanchnic mesoderm through its cognate receptor, FGFR2, in the lung endoderm. Acting as a chemotactic agent, FGF10 signaling pathway induces lung endodermal budding into the surrounding mesenchymal tissue. Deletion of FGF10 or its specific receptor, FGFR2, can strongly suppress early lung branching morphogenesis (reviewed by Warburton *et al.*, 2010; Morrisey and Hogan, 2010). Several signaling pathways are required for FGF10 expression specifically in the developing mesenchyme that surrounds the presumptive lung epithelium, including TGF-β, RA and Wnt/β-catenin signaling pathways (Chen *et al.*, 2010).

Sonic hedgehog (Shh) signaling pathway plays a major role in regulation of lung branching morphogenesis, while the Shh signaling downstream effectors, Gli2/3, control the early formation of lung buds (reviewed by Warburton *et al.*, 2010; Morrisey and Hogan, 2010).

However, the exact function and mechanism of action of Shh signaling to direct early development of the lung is still unclear, since Shh-null mutant mice still show a P–D patterning of the lung during embryogenesis. While FGF10 signaling pathway acts to direct the lung bud formation, a network of different signaling pathways such as Shh, RA, TGF-β and Wnt/β-catenin interact and integrate to regulate the proper formation and development of the embryonic lung.

Alternative populations of epithelial stem/progenitor cells exist in the distal lung. For example, alveolar stem cells that are integrin $\alpha_6\beta_4$-positive SFTPC-negative were identified and may contribute to the reconstitution of injured alveolar epithelial cells (Chapman *et al.*, 2011). In addition, murine distal airway stem cells may emerge or expand with remarkable reparative properties following a severe influenza infection. Remarkably, these distal airway stem cells express two genetic markers of the basal cells; TRP63 and KRT5, and can generate differentiated alveolar and bronchiolar epithelial cells after lung injury. In addition, another identified murine lineage-negative epithelial progenitor cell population can express KRT5 after activation and reconstitute both AEC2s and distal airway club cells after a severe influenza infection (Rawlins and Hogan, 2006; Rawlins *et al.*, 2008; Warburton *et al.*, 2010; Rock *et al.*, 2010; Rock and Hogan, 2011; Wansleeben *et al.*, 2013; Stabler and Morrisey, 2017). Notably, the differentiation of lineage-negative epithelial stem/progenitor cells into AEC2s is dependent on the suppression of notch signaling activity, whereas a persistent notch activity can result in the formation of cysts that are reminiscent of honey-combing in fibrotic human lungs (Vaughan *et al.*, 2015).

The balance of stem cell self-renewal/proliferation and differentiation acts to maintain homeostasis in different tissue types. Excess self-renewal of stem cells may result in tumorigenesis and/or tissue hyperplasia, while an abnormal increase of cell differentiation may cause degeneration and/or aging of the tissue. The asymmetric mode of stem cell division, producing one differentiating cell and one stem cell, is a simple and effective mechanism of maintaining the proper balance between differentiated cell populations and stem cell populations. Stem/progenitor cells normally undergo a mixture of both asymmetric and symmetric cell divisions. Generally, it is still difficult to clearly distinguish between these two different modes of cell division at the cellular level. A currently applicable rapid approach to distinguish between asymmetric and symmetric cell division is to detect differences in the mitotic spindle orientation or in the differential inheritance of cytoplasmic or membrane-bound cell fate determinants such as atypical PKC Zeta and Numb (Wang *et al.*, 2009; Huttner and Kosodo, 2005; Morrison and Kimble, 2006; El-Hashash and Warburton, 2011, 2012). Intrinsic or extrinsic fate determinant molecules play a key role in the asymmetric cell division. The cytoplasmic cell fate determinants such as Numb, which is an example of the intrinsic fate determinants, are localized asymmetrically within the cell. During cell division, these cell fate determinants normally segregate differentially into the two daughters; thus, the two daughter cells can take on two different fates. In case of the extrinsic fate determinant/micro-environment, the two daughters are grown in different microenvironments, and thus they take on different fates (reviewed in Yamashita, 2009, Berika *et al.*, 2014; Elshahawy *et al.*, 2016; El-Hashash, 2014; Ku and El-Hashash, 2016).

Balancing self-renewal with differentiation of stem/progenitor cells at lung distal epithelial tips is critical for normal lung development (El-Hashash and Warburton, 2011, 2012; El-Hashash *et al.*, 2011a; Berika *et al.*, 2014; Elshahawy *et al.*, 2016; El-Hashash, 2014; Ku and El-Hashash, 2016). Both cell polarity and the orientation of mitotic spindles play significant roles in the regulation of epithelial cell differentiation and self-renewal/proliferation. They can also impact many physiological and developmental processes such as epithelial cell differentiation and branching morphogenesis in different organs.

Understanding the behavior of lung-specific stem and progenitor cells and its regulatory molecular mechanisms could lead to the identification of innovative and important solutions to restore normal morphogenesis and function of the lung. However, little is still known about different aspects of lung stem cell behavior, spindle orientation, cell polarity and fate determination in the distal stem and progenitor cells of the embryonic lung. Distal epithelial stem/progenitor cells are characteristically polarized and highly mitotic as well as divide perpendicularly in the embryonic lung (El-Hashash and Warburton, 2011, 2012; Berika *et al.*, 2014; Elshahawy *et al.*, 2016; Ku and El-Hashash, 2016). In addition, Numb, which is well-known cell fate determinant, is distributed asymmetrically at the apical side of distal epithelial stem and progenitor cells of the embryonic lung. Numb normally segregates to one daughter of distal epithelial stem/ progenitor cells during mitosis, suggesting that these cells divide asymmetrically (ElHashash and Warburton, 2011, 2012; El-Hashash, 2014; Ku and El-Hashash, 2016; Elshahawy *et al.*, 2016).

5.3　Proximal–Distal Patterning of Lung Embryonic Cells

Both Bmp and Wnt signaling pathways play important roles in the regulation of the P–D patterning in the developing lung. Disruption of the activities of BMP or Wnt signaling pathways can lead to lung proximalization phenotypes (Li *et al.*, 2002; Eblaghie *et al.*, 2006). Whether the effects of BMP and Wnt signaling pathways are mediated through lung stem and progenitor cells still need more supportive evidences and studies. Indeed, the lung P–D patterning is controlled by Wnt/b-catenin signaling and is mediated at least partially by the regulation of several downstream targets such as N-myc, FGF and Bmp-4 signaling (Shu *et al.*, 2005). Moreover, in another study aimed to explore the functions of β-catenin-controlled gene expression in both maintaining and repair of lung bronchiolar epithelial cells using both cell type-specific knock-out and transgenic strategies, β-catenin was found to be unnecessary for maintaining adult bronchiolar stem cells (Zemke *et al.*, 2009). However, potentiation of the activity of β-catenin signaling can lead to an arrest of the differentiation of immature bronchiolar stem/progenitor cells (Zemke *et al.*, 2009).

Wnt5a also plays a critical role in both the morphogenesis of distal lung and the lung proximalization phenotypes (Li *et al.*, 2002). Mice carrying a targeted disruption of the Wnt5a locus show an abnormal truncation of the trachea and an overexpansion of the distal respiratory airways, indicating severe abnormalities in the morphogenesis of distal lung (Li *et al.*, 2002).

The study of the functions of Wnt signaling pathway during early stages of lung stem cell development is facilitate by the high activity levels of the reporters of Wnt signaling in distal lung epithelial stem/progenitor cells. Several studies have shown that Wnt signaling pathway can probably regulate the proliferation of lung stem and progenitor cells and lung P–D patterning independently. It is also important for the promotion of distal airway fate at the expense of the proximal airways (Shu *et al.*, 2005, Mucenski *et al.*, 2003). Indeed, inhibiting the activity of Wnt signaling by overexpressing the Wnt inhibitor Dickkopf-1 in the lung epithelial cells leads to the expansion of the proximal (conducting) airways at the expense of the distal airways, and without changing the total cell proliferation levels during lung development (Shu *et al.*, 2005). Moreover, β-catenin plays important roles in both Wnt signaling activity and adhesions between cells, and its lung-specific deletion leads to the inhibition of cell differentiation of distal airway epithelium (Mucenski *et al.*, 2003).

Notch signaling acts to promote stem/progenitor cell identity at the expense of differentiated cell phenotypes in many organ systems (Mizutani *et al.*, 2007; Jadhav *et al.*, 2006) and is an important regulator of lung epithelial stem/progenitor cells. Notch signaling is active in the distal epithelial stem/progenitors during the pseudoglandular stage and regulates the cell fate of the developing airways during embryogenesis (Post *et al.*, 2000; Tsao *et al.*, 2009). Misexpression of Notch1 signaling also acts to inhibit the differentiation of distal lung stem and progenitor cells before initiating of the alveolar program (Guseh *et al.*, 2009). Similarly, expressing active Notch3 can also inhibit cell differentiation in the developing lung epithelium (Dang *et al.*, 2003). In addition, active canonical Notch signaling pathway is important for both the selection of Clara versus ciliated cell fate and the arterial smooth muscle cell determination in the developing lung (Morimoto *et al.*, 2010).

Active BMP signaling is also important for the development of lung epithelium. BMP signaling may promote distal epithelial cell

fate but inhibit the proximal fate, as evidenced by the proximalization of the lung epithelium after inactivating BMP signaling by
the overexpression of the BMP antagonists Gremlin/Noggin, or a
dominant-negative BMP receptor (Weaver *et al.*, 1999; Lu *et al.*,
2001). However, the epithelial proximalization in the developing
lung is probably because of decreased activity of Wnt or BMP signaling pathway (Eblaghie *et al.*, 2006; Li *et al.*, 2002). In addition,
BMP4 can mediate the functions of other factors in the lung. For
example, the deficiency of histone deacetylases 1 and 2 (Hdac1/2)
in the lung can lead to a suppression of Sox2 expression and an
arrest of proximal airway development. This is partially mediated by
the expression of BMP4, a direct transcriptional target of Hdac1/2
(Wang *et al.*, 2013).

5.4 Role of Stem and Progenitor Cells in the Proximal–Distal Patterning of the Lung

During the growth of primary lung buds into the surrounding
splanchnic mesenchymal cells and their subsequent branching morphogenesis, many signals in the surrounding mesenchyme target lung
epithelial cells to control the P–D patterning of the lung. These regulatory signals from the neighboring distal mesenchymal cells are crucial for proper lung branching morphogenesis, as evidenced by early
tissue transplant experiments. In these experiments, the developing
respiratory epithelium during early embryogenesis (before E13.5) is
competent to properly respond to signaling pathways and molecules
from the adjacent distal mesenchyme, leading to promotion of airway
branching morphogenesis (Shannon and Hyatt, 2004).

Lineage-tracing experiments have used the Id2 locus to drive an
inducible cre/loxP system and provided evidences that lung stem and
progenitor cells exist in distal lung epithelial tips during branching
morphogenesis (Rawlins *et al.*, 2009). These distal stem/progenitor
cells can give rise to various types of epithelial cells during early lung
embryogenesis (before E13.5), suggesting the temporal multi-
potency nature of these cells (Rawlins *et al.*, 2009). Many signaling
molecules and pathways function to maintain the multi-potent nature

of lung distal stem and progenitor cells. For example, FGF signaling activities exist primarily at the lung distal branching tips and control the maintenance of the multi-potency of epithelial stem/progenitor cells, while active Wnt signaling functions to both maintain these epithelial progenitors and promote early distal epithelial identity prior to E14.5 (Warburton *et al.*, 2010; Morrisey and Hogan, 2010). Further studies are still need to identify and characterize signals and molecules that are important for the maintenance of lung stem/ progenitor cells that can help also in deriving stem/epithelial cells in the lung from pluripotent stem cells.

During lung branching morphogenesis, Sox2 is expressed in proximal lung epithelial cells and acts to promote proximal cell fate and suppress proximal cell branching (Que *et al.*, 2009). The proximal stem/progenitor cells that are localized in the lining of the epithelial stalks can generate different types of epithelial lineages that are localized in the upper airways such as ciliated cells and neuroendocrine (NE) cells as well as secretory Clara cells (Que *et al.*, 2009; Tompkins *et al.*, 2011). One of the crucial signaling pathways that acts to maintain the balance between secretory Clara cells and ciliated cells is Notch signaling. Indeed, Notch deficiency can lead to a loss of differentiation of secretory cells, and expansion of the ciliated cell lineage all over the proximal airways (Tsao *et al.*, 2009; Guseh *et al.*, 2009; Morimoto *et al.*, 2010).

Both Sox9 and Nkx2.1 transcription factors are characteristically expressed in distal lung epithelial stem/progenitor (Sox9+/Nkx2.1+) cells that can give rise to alveolar epithelial lineages such as type 1 (AEC1) and type 2 (AEC2) alveolar epithelial cells. In the murine lung, AEC1 and AEC2 lineages normally continue their maturation during the first several weeks of the post-natal development. This early maturation is necessary for organization of the growing and complex alveolar unit of gas exchange in the mature lung (Warburton *et al.*, 2010). A characteristic network of interactions between AEC2 cells, AEC1 cells and neighboring vascular, endothelial and mesenchymal fibroblast cells is important to promote the proper and normal maturation of the developing alveolar unit. Proper control of the maturation of the developing alveolar unit is crucial for normal lung development and function since any deficiency in this important process can lead to several human pulmonary diseases such as alveolar capillary dysplasia and bronchopulmonary dysplasia (BPD) (Warburton *et al.*, 2010).

5.5 Summary and Conclusions

During embryonic development, the endoderm represents the inner germ layer in both triploblastic and diploblastic organisms and is responsible for the formation of the internal epithelial tube that will eventually form the digestive tract. The endodermal germ layer is, therefore, a major contributor for the formation of both the gastro-intestinal and respiratory tracts, and their associated organs. Both the P–D and A–P patterning are among the earliest events during embryonic development. Both P–D and A–P patterning are processes tightly regulated by several molecular mechanisms and signaling pathways such as FGF, BMP, Shh, retinoic acid and Wnt signaling pathways, as well as serval transcription factor families, including Sox, Hhex, Pdx1 and Cdx.

Several molecular mechanisms and signaling pathways have been implicated in the control of early foregut specification through well-investigated endodermal–mesodermal signaling interactions, including FGF, BMP, Notch, Wnt, Shh, FGF and RA, while members of the TGF-β signaling network such as Nodal are essential regulators of the early development of a definitive endoderm.

Furthermore, many molecular mechanisms and signaling molecules from the surrounding mesenchymal cells are major regulators of the foregut patterning along the D–V axis. In addition, the D–V patterning of foregut endodermal stem and progenitor cells is tightly regulated by a network of signaling molecules and pathways from the neighboring mesenchymal cells like the A–P patterning. The BMP signaling pathway, for example, is one of these signaling pathways that has spatial-specific and temporal functional roles in the regulation of how the foregut endodermal cells specify and differentiate. In addition, BMP signaling activities from the ventral mesenchyme to the developing endoderm regulate the D–V patterning of the murine anterior foregut during early developmental stages.

Signaling pathways are also important in the development and patterning of endodermal derivatives such as the lung. During the primary lung bud growth into the surrounding splanchnic mesenchymal cells and their subsequent branching morphogenesis, many signals in the surrounding mesenchyme target lung epithelial cells to control the P–D patterning of the lung. These regulatory signals from the neighboring distal mesenchymal cells are also essential for proper lung branching morphogenesis, as evidenced by early tissue transplant experiments. Indeed, a well-coordinated network of different signaling pathways such as TGF-β, RA, Shh, FGF10 and Wnt/β-catenin interact and integrate to regulate the proper formation and development of the embryonic lung. Shh signaling pathway, for example, interacts with FGF10 signaling to control branching morphogenesis in the lung, while Gli2/3, which are Shh signaling downstream effectors, regulate the early lung bud formation. Since Shh mutant mouse embryos still show a P–D patterning of the lung, more research is still needed to determine the exact functions and mechanism of actions of Shh signaling in the early lung development during embryogenesis.

Furthermore, both Wnt and BMP signaling pathways regulate the P–D patterning in the developing lung. Thus, the Wnt/b-catenin signaling controls the lung P–D patterning through targeting the activities of several downstream targets such as N-myc, and both FGF and BMP-4 signaling elements, and is important for the promotion of distal airway fate at the expense of the proximal airways. For example, Wnt5a regulates both the morphogenesis of distal lung and the lung proximalization phenotypes. Beside the P–D patterning, the Wnt signaling pathway can also control the proliferation of lung stem and progenitor cells. However, it controls these two processes independently. Similarly, active BMP signaling regulates the development and patterning of lung epithelial cells by promoting distal epithelial cell fate but inhibiting the proximal fate, a finding that is supported by the proximalization of the lung epithelium after inactivating BMP signaling through overexpressing the BMP antagonists Gremlin/Noggin, or a dominant-negative BMP receptor.

Chapter 6

Lung Cell Polarity, Fate and Mode of Division

Abstract

New data have recently accumulated on how stem cell behave, self-renew and differentiate. Many studies have also focused on defining stem cells, and determination of the properties, including the mode of cell division and polarity, and regulatory environment(s) of both embryonic and tissue-specific stem cells in the last decades. In the lung, recent data show evidences that lung epithelial stem and progenitor cells are polarized, highly mitotic, have characteristic perpendicular cell divisions, and show a mode of division that is similar to other systems. They further show that the asymmetric division is probably the common mode of division in the mitotically dividing distal epithelial stem and progenitor cells of the embryonic lung. Both symmetric and asymmetric mode of cell divisions are tightly regulated in different stem cell types during tissue development and morphogenesis. How to choose between a symmetric and asymmetric cell division is one of the major questions in the stem cell field. It largely affects tissue development, morphogenesis and disease in different organs since improper asymmetric divisions badly affect organ morphogenesis, whereas uncontrolled symmetric division can lead to tumor formation. Moreover, the proper balance between self-renewal and differentiation of lung epithelial stem and progenitor cells is absolutely required for maintaining normal lung morphogenesis and for lung repair and regeneration since a deficiency of this balance probably can lead to a premature or injured lung. Therefore, identification of lung-specific stem cell types, understanding their behavior, and how they balance their self-renewal and differentiation could lead to the identification of innovative solutions for restoring normal lung morphogenesis and/or regeneration and repair of the lung. Furthermore, understanding the molecular mechanisms that control the asymmetrical cell division and both cell polarity and fate of lung epithelial stem and progenitor cells can help identifying new targets for prevention and rescuing lethal lung diseases in infants and children, and for regeneration of injured lungs. In this chapter, we will discuss recently accumulated data on the lung cell polarity, and the mode of division of lung epithelial stem and progenitor cells. In addition, we will describe the functions of Numb in stem cell fate and mode of division, and compare cell polarity and mode of division in the lung stem cells with other systems, as well as discuss the regulatory mechanisms of lung stem cell polarity, fate, behavior and mode of division.

Keywords: Lung; cell polarity; cell division; stem and progenitor cells; Numb; cell fate; asymmetric cell division; cell behavior; transcription factors; growth factors; signaling pathways.

6.1 Lung Cell Polarity

The cell polarity is dependent on the asymmetric distribution of the cellular components inside the cell. Both the cell polarity and orientation of mitotic spindle have pivotal roles in the self-renewal/proliferation and differentiation of epithelial cells and can affect many physiological processes, such as epithelial differentiation and branching morphogenesis. In addition, cell polarity can help in organizing and integrating complex molecular signaling in the epithelial cells. Therefore, these cells can decide their fate and whether to proliferate or differentiate (Drubin and Nelson, 1996; Martin-Belmonte and Perez-Moreno, 2011; El-Hashash and Warburton, 2011; Kim *et al.*, 2018).

Distal epithelial stem and progenitor cells are polarized, highly mitotic and divide perpendicularly in the embryonic lung (El-Hashash and Warburton, 2011, 2012; Berika *et al.*, 2014; Ibrahim and El-Hashash, 2015; Elshahawy *et al.*, 2016). In addition, cell polarization plays an important role in the perpendicular cell division of distal epithelial stem and progenitor cells in the lung (Wodarz, 2002; Elshahawy *et al.*, 2016). Disruption of cell polarization can lead to loss of the balance between self-renewal and differentiation in lung epithelial cells (El-Hashash and Warburton, 2011, 2012; Berika *et al.*, 2014; Ibrahim and El-Hashash, 2015). A correlation between the perpendicular cell division and asymmetric division has been reported in the epithelial cells of different tissues in mammals (Lechler and Fuchs, 2005). This was also confirmed in the embryonic lung epithelium by detecting the asymmetric localization of proteins controlling the mitotic spindle orientation, including nuclear mitotic apparatus (NuMA), mouse Inscuteable (mInsc) and G-protein signaling modulator 2 (Gpsm2), polarity proteins in mitotic distal lung epithelial stem and progenitor cells (El-Hashash and Warburton, 2011, 2012; Berika *et al.*, 2014; Ibrahim and El-Hashash, 2015).

The epithelial cell has a characteristic apical–basal polarity in different organs, such that the switch of epithelial cells from symmetric to asymmetric cell division occurs mostly with only subtle deviation of the spindle orientation that can result in the asymmetrical distribution of their adherent junctions and apical plasma membrane to the daughter cells (Nelson, 2003; Kosodo *et al.*, 2004). E-cadherin is a major component of the lateral epithelial cell plasma membrane and the apico-lateral junction complex (Woods *et al.*, 1997). The plasma membrane of mitotically dividing cells shows the "cadherin hole" as a comparatively small unstained segment in the cell surface by immunostaining for E-cadherin (El-Hashash and Warburton, 2012; Berika *et al.*, 2014; Ibrahim and El-Hashash, 2015). Furthermore, the cleavage plane orientation relative to the cadherin hole can predict whether symmetric or asymmetric distribution of the plasma membrane to daughter cells occurs, which is an indication of whether the mitotic cell divides symmetrically or asymmetrically in the epithelium of different organs (Kosodo *et al.*, 2004). Remarkably, the analysis of cadherin hole in embryonic lung epithelial cells demonstrates that most distal epithelial stem/progenitor cells divide asymmetrically (El-Hashash and Warburton, 2012; Berika *et al.*, 2014; Ibrahim and El-Hashash, 2015). The cleavage plane of the mitotically dividing cell is supposed to circumvent the cadherin hole to the daughter cell that further supports the asymmetric division of most distal lung epithelial stem and progenitor cells in the embryonic lung (El-Hashash and Warburton, 2012; Berika *et al.*, 2014; Ibrahim and El-Hashash, 2015).

6.2 Modes of Lung Stem and Progenitor Cell Division

Proper regulation of epithelial stem and progenitor cells is crucial for development of mammalian lung (Warburton *et al.*, 2010; Warburton *et al.*, 2008). Congenital defects of lung stem and progenitor cells can lead to many fatal abnormalities or disorders such as bronchopulmonary dysplasia (BPD) or lung hypoplasia by affecting vital biological processes, including the capacity of gas diffusion (Warburton *et al.*, 2010; Warburton *et al.*, 2008; Shi *et al.*, 2009; Warburton *et al.*, 2000). The balance between stem cell self-renewal and differentiation is one of the major mechanisms of maintaining tissue homeostasis in different organs including lung. Tissue hyperplasia and/or tumorigenesis may result from an excessive self-renewal of stem cells; while unregulated increases of cell differentiation may lead to tissue degeneration and/or aging (Warburton *et al.*, 2010; Berika *et al.*, 2014; Ibrahim and El-Hashash, 2015). Understanding the mechanism of proper balance between lung stem/progenitor cell self-renewal and differentiation is, therefore, critical for the identification of new solutions to repair the gas diffusion surface and restore lung morphogenesis. During lung development, asymmetric cell division is essential for the proper balance between stem cell self-renewal and differentiation and to correct temporal and spatial specification of epithelial cell lineages (Knoblich, 2001; Yamashita *et al.*, 2010; Berika *et al.*, 2014; Ibrahim and El-Hashash, 2015; Elshahawy *et al.*, 2016).

The growth of stem and progenitor cell population in the lung is dependent on the mode of their cell division (symmetric or asymmetric; Lu *et al.*, 2008; Rawlins, 2008). There are few feasible ways to differentiate between symmetric and asymmetric mode of division, including observing both the mitotic spindle orientation and differences in the inheritance of the membrane-bound or cytoplasmic proteins such as the cell fate determinant Numb and atypical PKCζ (Huttner and Kosodo, 2005; Morrison and Kimble, 2006; Wang *et al.*, 2009; El-Hashash and Warburton, 2011, 2012). The asymmetrical mode of stem cell division is dependent on extrinsic or intrinsic cell fate determinants. The stem cell micro-environment is an example of extrinsic fate determinants. If each daughter of a dividing stem cell is placed in different micro-environment, it will acquire a different fate. The well-investigated cytoplasmic cell fate determinant Numb is an example of intrinsic cell fate determinant. The asymmetrical mode of stem cell division is largely dependent on the preferential segregation and inheritance of certain intrinsic cell fate determinant molecules/factors, including Numb, into one daughter of the dividing stem/progenitor cell in the epithelium of *Drosophila* and mammals (Betschinger and Knoblich, 2004; Cayouette and Raff, 2002). Both the asymmetric distribution of cell fate determinants in the dividing cell and defining the axis of polarity can influence the apical–basal polarity and allow a rapid switch from proliferation to differentiation in the epithelial cell (Betschinger and Knoblich, 2004).

The cellular distribution of some other cell components such as E-cadherin can also help in predicting the mode of cell division. E-cadherin is a component of the apico-lateral junctional complex and a part of the lateral epithelial cell plasma membrane (Woods *et al.*, 1997). The immunostaining for E-cadherin in the plasma membrane of dividing epithelial cells during mitosis shows an unstained segment that is called the cadherin hole (Kosodo *et al.*, 2004; El-Hashash and Warburton, 2012). Early studies have indicated that the symmetric vs. asymmetric distribution of the plasma membrane to daughter cells can be predicted from the orientation of the cleavage plane relative to the cadherin hole of the mitotic epithelial cell in different organs (Kosodo *et al.* 2004). Cadherin hole analysis studies have demonstrated that the alteration in the orientation of the cleavage plane in distal lung epithelial stem and progenitor cells is correlated with the distribution of the cell membrane/cadherin hole during cell division (El-Hashash and Warburton, 2012). Thus, in distal lung epithelial stem and progenitor cells, most (94%) of the cleavage planes are predicted to bypass the cadherin hole, resulting in the asymmetric distribution of the cadherin hole to the daughter cells, whereas a small percentage (6%) of the cleavage plane orientations are predicted to bisect the cadherin hole, which results in a symmetric distribution to both daughter cells. This analysis of the cadherin hole in the lung epithelium provides evidences that the asymmetric division is common in distal epithelial stem and progenitor cells (El-Hashash and Warburton, 2012; Berika *et al.*, 2014; Elshahawy *et al.*, 2016).

6.3 Numb and Lung Stem Cell Fate and Mode of Division

The preferential segregation of the intrinsic cell fate determinants functions to mediate the asymmetric mode of cell division in *Drosophila* and various types of mammalian epithelial cells. Early studies have shown that the functions of the cell fate determinants are to define the axis of polarity that determines the orientation of the apical–basal cell plane for division (Betschinger and Knoblich, 2004). The cell fate determinants are asymmetrically localized in the cells to switch between proliferation/self-renewal (i.e. two similar daughter cells are produced) and diversification (i.e. different-shaped daughter cells are produced; Betschinger and Knoblich, 2004). Therefore, the asymmetrical localization of well-characterized determinants of stem cell fate (e.g. Numb) and definition of the axis of polarity in the dividing cells can determine their apical–basal orientation and enable them to rapidly switch from proliferation to differentiation (Betschinger and Knoblich, 2004).

Numb is one of the most studied cell fate determinants and is critical for the determination of the mode of cell division, asymmetric vs. symmetric cell division, of epithelial cells in different tissue types (Morrison and Kimble, 2006; El-Hashash and Warburton, 2011; El-Hashash and Warburton, 2012; Zimdahl *et al.*, 2014). Numb protein is expressed uniformly in the cell cytoplasm during interphase. Numb has an asymmetrical distribution and localization in the cytoplasm during cell division, and it segregates to one of the two daughters in most mitotically dividing epithelial cells. Since Numb acts to inhibit the Notch signaling activity that maintains the stem cell identity in different systems/tissue types, the two daughter cells will acquire different fates based on their inheritance of Numb that influence the Notch activity. The daughter cell with high Numb levels shows a suppression of Notch signaling activity and differentiates, while the other daughter receiving low Numb levels can preserves high Notch activity and, therefore, acquire a stem cell fate (Frise and Knoblich, 1996; Frise *et al.*, 1996; Guo *et al.*, 1996; Juven-Gershon *et al.*, 1998; Betschinger and Knoblich, 2004; Alexson *et al.*, 2006; Yan *et al.*, 2008; El-Hashash and Warburton, 2011, 2012; Berika *et al.*, 2014; Elshahawy *et al.*, 2016).

In the embryonic lung, Numb is asymmetrically distributed and highly expressed on the apical side of distal lung epithelial stem and progenitor cells, in which it plays a critical role in the mode of cell division: asymmetric vs. symmetric division (El-Hashash and Warburton, 2011, 2012; El-Hashash and Warburton, 2011, 2012; Berika *et al.*, 2014; Elshahawy *et al.*, 2016). In addition, Numb protein segregates to and is inherited by one daughter cell in most mitotic distal lung epithelial stem and progenitor cells (El-Hashash and Warburton, 2011, 2012; Berika *et al.*, 2014; Elshahawy *et al.*, 2016). The asymmetric segregation and inheritance of Numb by one daughter cell in dividing distal lung epithelial stem and progenitor cells provide another evidence that these cells undergo asymmetric cell divisions (El-Hashash and Warburton, 2012). In addition, there is a significant increase of cells that express Sox9/Id2 (epithelial stem cells markers) when Numb is knocked down in MLE15 lung epithelial cells in culture, which supports Numb function as a cell fate determinant in the lung (El-Hashash and Warburton, 2012).

6.4 Similarity of the Cell Polarity, Fate and Mode of Division in Lung Stem Cells and Other Systems

There is a similarity in the mode of division of stem and progenitor cells between the lung and other tissues or systems. For example, distal lung epithelial stem and progenitor cells are polarized with preferential perpendicular rather than parallel cell divisions (El-Hashash and Warburton, 2011; Berika *et al.*, 2014; Elshahawy *et al.*, 2016), like stem and progenitor cells of other types of tissues (Lechler and Fuchs, 2005; Yamashita *et al.*, 2010). In addition, both the asymmetric segregation and inheritance of the cell fate determinant Numb may be the most common mode of controlling the asymmetric cell division in lung epithelial stem and progenitor cells (El-Hashash and Warburton, 2012; Berika *et al.*, 2014; Elshahawy *et al.*, 2016), similar to neural stem and progenitor cells and satellite muscle cells (reviewed by Morrison and Kimble, 2006).

Another notable aspect of similarity between the lung epithelial stem cells and other stem cell types is the asymmetric apical localization of Par/LGN/NuMA/mInsc polarity proteins, which also control the mitotic spindle orientation in various types of mitotic epithelial cells in mammals (Lechler and Fuchs, 2005). Indeed, the most distal epithelial stem and progenitor cells have apically localized polarity proteins such as Par, LGN, NuMA and mInsc, with mitotic spindles aligned perpendicular to the basement membrane in the lung. Consistently, several studies have shown that both the polarized localization of Par, LGN, NuMA and mInsc proteins and the perpendicular alignment of mitotic spindles are strictly correlated with the asymmetric mode of cell division in *Drosophila* and mammalian epithelial cell types (Cayouette and Raff, 2002, 2003; Haydar *et al.*, 2003; Noctor *et al.*, 2004; Lechler and Fuchs, 2005).

The asymmetric mode of cell division mediates the balance between stem cell self-renewal and differentiation in different types of tissues and systems (Yamashita *et al.*, 2010). This asymmetric cell division-mediated balance is essential for the long-term maintenance of tissue self-renewal during development, repair and regeneration as well as in diseases in different types of organs, including the lung. For example, both congenital lung hypoplasia and BPD, wherein a significant deficiency of lung stem cells probably occurs, are common features of human prematurity and/or lung injury and are, thus, major public health problems in human infancy. The proper balance between self-renewal and differentiation of lung-specific stem and progenitor cells, which is mediated by the asymmetric mode of cell division is, therefore, most likely required for normal lung development and for both lung repair and regeneration after injury. Remarkably, the tightly controlled outgrowth and branching morphogenesis of the epithelial tubes in the developing lung generate a sufficiently large gas diffusion surface to sustain life. The developmental disorders or defects that affect this smooth progression may, therefore, lead to a defective differentiation and severe diseases such as postnatal respiratory distress (Warburton *et al.*, 2008, 2010). The similarities of the asymmetric stem and progenitor cell division between the lung and other tissue systems, as reported in our laboratory, could identify innovative solutions for restoring normal lung morphogenesis and for lung repair and regeneration after injury.

6.5 Regulatory Mechanisms of Lung Stem Cell Polarity, Fate, Behavior and Mode of Division

Many transcription factors and signaling molecules were reported to control the expression of the regulatory genes of stem and progenitor cell self-renewal, proliferation and differentiation, and therefore play critical roles in lung development, morphogenesis, repair and regeneration. For instance, Ttf-1 transcription factor plays a major role in the development of distal lung stem and progenitor cells and in the lineage commitment in the early embryonic lung (Kimura *et al.*, 1999). Ttf-1-mutant mice are born dead owing to remarkable abnormalities in cell differentiation in the lung (Kimura *et al.*, 1996). In addition, the HMG transcription factor Sox9 is highly expressed in the distal epithelial stem and progenitor cells in embryonic lung.

However, conditional deletion of Sox9 in the embryonic lung does not apparently affect the behavior of stem and progenitor cells in the lung (Perl *et al.*, 2005; Liu and Hogan, 2002; Rockich *et al.*, 2013). Other genes may act concomitant with Sox9 to regulate stem and progenitor cell behavior and, therefore, can compensate for the loss of Sox9 functions in the lung. For example, the overexpression of the transcription factor N-myc can result in the inhibition of lung stem/progenitor cell differentiation, in addition to an increased expression of Sox9 in the distal lung epithelial compartments (Okubo *et al.*, 2005).

Furthermore, forkhead box (fox) transcription factors are critical regulators of stem and progenitor cell self-renewing divisions during lung morphogenesis. Fox family members include Foxa1, Foxa2, Foxj1, Foxf1, Foxp1 and Foxp2, and these may control the expression of many cell proliferation regulatory genes in the developing lung. Consequently, a conditional deletion of both *foxa1* and *foxa2* genes can markedly decrease cell proliferation in the lung, leading to a hypoplastic lung phenotype (Wan *et al.*, 2005). Furthermore, a conditional deletion of both foxp1 and foxp2 genes *can* also result in a similar lung phenotype to that found in *foxa1* and *foxa2-* deficient mice. Thus, *foxp22/2; foxp11/2* double mutant mice show lungs that are small with a remarkable decrease in cell proliferation, but have a regular proximal–distal (P–D) patterning (Shu *et al.*, 2007). In addition, the mesenchymal nuclear factor 1 B (NF1B) is a transcription factor that can regulate epithelial cell self-renewal and differentiation during maturation of the lung (Hsu *et al.*, 2011).

Other transcription factors regulate cell proliferation in the lung. For example, MyoD is a muscle-specific helix–loop–helix transcription factor that has essential roles in both the proliferation and differentiation of myoblast cells in different organs (Tapscott, 2005; Berkes and Tapscott, 2005; Bryson-Richardson and Currie, 2008; Bentzinger *et al.*, 2012). MyoD−/− mutant lungs show a decrease of cell proliferation at E18.5 and are, therefore, hypoplastic (Inanlou and Kablar, 2003). In addition, in MyoD−/− mice diaphragm muscle cannot support fetal breathing movements because it is significantly thin (Inanlou and Kablar, 2003).

There are five major signaling pathways that regulate the embryonic development: Fibroblast growth factor (FGF), transforming growth factor-b (TGF-b) family, Wnt and hedgehog as well as Notch pathways. These signaling pathways also play important functional roles in the specification, self-renewal, proliferation and differentiation of different types of stem and progenitor cells. The current consensus is that these signaling mechanisms collaborate to regulate the proliferation and differentiation of distal epithelial stem and progenitor cells in the lung. Despite their importance, little is known about the mechanisms of these signaling molecules in lung stem and progenitor cells and how they interact.

Growth factors play essential functional roles in the embryonic lung development. FGF signaling pathway has a major functional role in specification of the lung lineages distal to the trachea (Ramasamy *et al.*, 2007; Serls *et al.*, 2005; El Agha *et al.*, 2014). FGF10, for example, is expressed by lung mesenchymal cells and plays a pivotal function in the amplification of lug epithelial stem and progenitor cells (Ramasamy *et al.*, 2007). FGF10 signaling can also organize alveolar smooth muscle cell development and angiogenesis (Ramasamy *et al.*, 2007). In addition, the overexpression of FGF10 can increase the proliferation of epithelial progenitor cells and lead to hyperplasia of goblet cells (Nyeng *et al.*, 2008). FGF10 signaling activities are regulated by several factors in the lung, including the retinoic acid and the RNA-binding protein HuR. Retinoic acid signaling has an important role in both the proliferation of lung stem and progenitor cells and development of primary lung buds, by influencing FGF10 expression through TGF-beta signaling pathway (Chen *et al.*, 2007). In addition, HuR is an RNA-binding protein (RBP) that controls the mesenchymal responses during lung branching morphogenesis (Sgantzis *et al.*, 2011). HuR regulates the mRNA levels Fgf10 and Tbx4. Deletion of HuR can diminish the post-transcriptional regulation of Fgf10 and Tbx4, leading to a severe reduction of distal bronchial branching morphogenesis during the early pseudoglandular stage of lung development (Sgantzis *et al.*, 2011).

BMP signaling is a member of the TGF-β superfamily and is essential for the proliferation of lung distal epithelial stem and progenitor cells. Thus, the autocrine signaling through BMP receptor 1 (Bmpr1) can control cell division and survival of murine distal lung epithelial cells during lung morphogenesis (Eblaghie *et al.*, 2006). The deletion of either Bmpr1a receptor in the lung epithelium or its ligand, Bmp4, can lead to a hypoplastic lung phenotype, with a remarkable reduction of both cell proliferation and expression levels of N-myc and foxa2 (Eblaghie *et al.*, 2006).

Canonical Wnt2/2b and beta-catenin signaling can play essential roles in the specification of lung endoderm progenitor cells in the growing foregut endoderm (Goss *et al.*, 2009). Null alleles of both *Wnt2* and *Wnt2b* genes were produced in murine embryos using homologous recombination. These murine embryos show a remarkable lung agenesis with a complete absence of tracheal budding (Goss *et al.*, 2009). Notably, a conditional inactivation or activation of *beta-Catenin* gene can result in the loss or gain of trachea/lung progenitor cell identity in murine foregut endoderm, respectively (Goss *et al.*, 2009; Harris-Johnson *et al.*, 2009). Similarly, Wnt5a, which is a prominent Wnt ligand, is intensively expressed in and around the distal epithelial tips of the developing lung. Conditional deletion of Wnt in mice can lead to a significant increase in cell proliferation, leading to the formation of additional branches of conducting airways (Li *et al.*, 2002). However, further studies are still needed to know whether this phenotype is related to defects in stem and progenitor cells.

The hedgehog signaling pathway functions to maintain the epithelial quiescence in different body systems. Sonic hedgehog (Shh) signaling plays an essential role in both the epithelial cell proliferation and branching morphogenesis in the developing lung (Pepicelli *et al.*, 1998). Indeed, the suppression of hedgehog signaling activity can result in an increase of the epithelial cell proliferation at homeostasis and after lung injury (Peng *et al.*, 2015; Metcalfe and Siebel, 2015). In addition, the overexpression of Gli2, a key mediator of SHH activity, is associated with increased cell proliferation and expression of cyclin proteins that regulate the progression of cells through the cell cycle in the lung. Thus, the Gli2-mediated SHH signaling may regulate the development of embryonic lung through targeting cyclin proteins (Rutter *et al.*, 2010). On the other hand, the canonical Notch signaling is essential for the determination of Clara vs. ciliated cell fate and for the identification of arterial smooth muscle cells in the developing lung (Morimoto *et al.*, 2010).

Different other factors regulate the behavior, polarity, mode of division and fate of progenitor cells in the lung such as miRNAs and protein phosphatases. Micro-RNAs (miRNAs) are a class of small, non-coding RNAs that have the capability to regulate gene expression and act as a post-transcriptional modifier (Huang *et al.*, 2011; Alberti and Cochella, 2017). The importance of miRNA in lung cell proliferation and differentiation has been demonstrated in several reports. Thus, the over-expression of miR-17–92 in embryonic lung epithelial cells results in increased number of highly proliferative, undifferentiated epithelial cells that express the progenitor cell marker Sox9, suggesting that miR-17–92 stimulates the proliferation of lung epithelial progenitor cells (Lu *et al.*, 2007). During the pseudoglandular stage of lung development, both miR-17 and its paralogs, miR-20a and miR-106b, are highly expressed in the lung, and important for the embryonic lung growth and development (Carraro *et al.*, 2009). In addition, concurrent down-regulation of these three miRNAs in isolated lung epithelial explants grown in culture can modify and change FGF-10-induced budding morphogenesis, an effect that could be rescued by expressing synthetic miR-17 in these explants (Carraro *et al.*, 2009). In addition, down-regulation of these three miRNAs leads to the inhibition of E-Cadherin expression levels and disruption of its distribution in the cell. However, the activity of beta-catenin activity increases, and expression of its downstream targets, BMP4 and Fgfr2b receptor, is augmented (Carraro *et al.*, 2009).

Furthermore, both signal transducer and activator of transcription 3 (Stat3) and mitogen-activated protein kinase 14 (Mapk14) are targets of the activity of miR-17, miR-20a and miR-106b in the developing lung (Carraro *et al.*, 2009). Notably, concurrent overexpression of Stat3 and Mapk14 can rescue changes in E-Cadherin distribution that occur after the co-down-regulation of miR-17, miR-20a and miR-106b in lung epithelial expants grown in culture, indicating that mir-17 family of miRNA alters FGF10-FgfR2b downstream signaling by distinctively targeting Stat3 and Mapk14, thus controlling the expression of E-Cadherin. This consecutively alters the morphogenesis of lung epithelial buds in response to FGF-10 signaling (Carraro *et al.*, 2009).

Protein phosphatases play important roles in the regulation of cell polarity, fate and mode of division in the lung epithelium. Eya1 protein phosphatase, for example, is highly expressed in distal lung epithelial stem and progenitor cells and controls their polarity, mitotic spindle orientation and Numb distribution patterns (Berika *et al.*, 2014; Elshahawy *et al.*, 2016). Eya1 protein phosphatase targets and controls the phosphorylation and activity of aPKCζ, which is a well-known regulator of both the perpendicular divisions and the asymmetrical Numb inheritance in dividing cells (Berika *et al.*, 2014; Elshahawy *et al.*, 2016; Tepass, 2012; Drummond and Prehoda, 2016). Thus, Eya1 promotes the perpendicular cell division and plays a critical functional role in the asymmetrical segregation and inheritance of Numb during mitosis in the dividing distal lung epithelial stem and progenitor cells, probably by the regulation of the phosphorylation and activity levels of aPKCζ (Berika *et al.*, 2014; Elshahawy *et al.*, 2016).

6.6 Summary and Conclusions

Recent studies show that distal lung epithelial stem and progenitor cells are polarized, highly mitotic and have characteristic perpendicular cell divisions. The disturbance of cell polarity in lung stem or progenitor cells can lead to loss of the balance between self-renewal and differentiation. In addition, other studies have shown that asymmetric division is common in the mitotically dividing distal epithelial stem and progenitor cells of the embryonic lung. This conclusion was supported with several evidences. For example, the orientation of the cleavage plane was shown to circumvent the cadherin hole, leading to the asymmetric distribution of cadherin hole between daughter cells of the mitotically dividing embryonic distal lung epithelial stem and progenitor cell. Moreover, most of distal lung epithelial stem and progenitor cells contain apically localized polarity proteins such as LGN, Par, NuMA and mInsc that control the mitotic spindle orientation in dividing cells. They also show asymmetric segregation and inheritance of Numb during their mitotic division that is another evidence for the asymmetric mode of cell division. Moreover, the mitotic spindle of these lung epithelial stem and progenitor cells are aligned perpendicular to the basement membrane.

Furthermore, the proper balance between self-renewal and differentiation of lung epithelial stem and progenitor cells is absolutely required for maintaining normal lung morphogenesis and for lung repair/regeneration since a deficiency of this balance probably can lead to a premature or injured lung. Therefore, identifying the molecular mechanisms that regulate the balance between self-renewal/ proliferation and differentiation of distal lung epithelial stem and progenitor cells is essential for proper lung development and morphogenesis as well as for lung repair and regeneration. In addition, the proper outgrowth and branching of the epithelial cells is crucial for generating a large enough gas diffusion surface in the lung. Thus, developmental deformity in this smooth progression of the balance between lung stem cell self-renewal and differentiation can lead to improper differentiation and possible postnatal respiratory distress. Better understanding of lung stem cell behavior may, therefore, lead to identifying novel treatments for abnormal lung morphogenesis.

Future studies that focus on understanding the molecular and cellular mechanisms that control the asymmetric cell division and cell polarity as well as the balance between self-renewal and differentiation of lung epithelial stem and progenitor cells can help identifying new targets for prevention and rescuing lethal lung diseases in infants and children and for regeneration of injured lungs.

Chapter 7

General Summary, Concluding Remarks and Future Directions

The adult lung has a complex structure that is composed of multiple specialized epithelial cell types, fibroblasts, two distinct smooth muscle populations (bronchial and vascular) and two parallel circulations (systemic and pulmonary), as well as a unique local immune system. This complexity of the lung structure is one of the major factors that has hampered the investigation of molecular and cellular mechanisms controlling lung cell development, behavior and function during both homeostasis and repair. However, the recent development of novel lineage-tracing tools and the discovery of informative cell markers have facilitated the identification of lung progenitor cell hierarchies.

The common consensus is that the epithelial stem cells leave the progenitor pool to produce lineage-committed precursor cells that in turn give rise to fully differentiated cell lineages. Rapid and controlled self-renewal of lung stem and progenitor cells and differentiation of these cells can produce sufficient large alveolar gas diffusion surface to maintain normal post-natal life. Abnormalities or defects in this balanced progression of lung development may lead to abnormal cell differentiation, and therefore post-natal respiratory distress in humans (Warburton *et al.*, 2010; Berika *et al.*, 2014; El-Hashash, 2013, 2014; Ibrahim and El-Hashash, 2015; Elshahawy *et al.*, 2016).

Stem studies using cells have suggested several mechanisms for the production and maintenance of new stem cells as well as generation of differentiated cells and focused on the asymmetric cell division in different organs and species. For example, *Drosophila* is a well-known invertebrate model that is used in the discovery of the functional role of different extrinsic signals and intrinsic factors in the control of stem cell division and how both these signals and factors act together to coordinate and control the asymmetric divisions. Several other studies show similar mechanisms using *in vitro* models. However, the characterization of stem cells *in vivo* and identification of the molecular mechanisms that regulate asymmetric cell division in mammals, including humans, still need more advances and improvements in the research techniques such as the isolation and purification of stem cells and real-time imaging of these cells. An increase of symmetric cell divisions may be temporarily needed to provide the body tissue with enough number of stem and progenitor cells during the processes of tissue formation, regeneration and repair after injury. Understanding the mechanisms that control asymmetric cell division is, therefore, critical for organ development as well as during tissue repair and regeneration after injury. In addition, it is important to identify and characterize factors and molecular mechanisms that act to retard or even prohibit stem/progenitor cells from switching from symmetric mode of cell division back to asymmetric cell division. A chronic inflammation or damage of a tissue represents a good example of these factors since they might modify the ability of stem and progenitor cells to respond effectively to repair the injured tissue. In addition, tissue damage may prevent stem and progenitor cells from switching from symmetric to asymmetric mode of cell division.

Remarkably, the inappropriate regulation of tissue repair may eventually lead to choosing stem and progenitor cells that can resist signals controlling the normal growth, which is a hallmark of cancer cells. Thus, identifying and characterizing the signaling mechanisms that control the asymmetrical mode of cell division in different types of stem cells is important for developing approaches that exploit the ability of these stem cells to repair diseased or injured body tissues. In addition, better understanding of these molecular mechanisms can lead to designing new strategies to suppress cancer development in different cell types and may also lead to the development of novel targets for anti-cancer therapies. Furthermore, understanding the molecular mechanisms and signals regulating the mode of proliferation or differentiation in adult stem cells will provide new strategies for the maintenance and expansion of stem cells *in vitro*, while preserving their differentiation potential. These novel strategies and approaches can also help in directing the differentiation of stem cells into various cell types that can be utilized in regenerative medicine.

The mode of stem and progenitor cell division in the lung is similar to that in other organs. For example, most stem and progenitor cells in lung distal epithelium are polarized and prefer a perpendicular and asymmetric division rather than parallel and symmetric division, similar to many other tissues (El-Hashash and Warburton, 2011; El-Hashash and Warburton, 2012; Berika *et al.*, 2014; Elshahawy *et al.*, 2016). In addition, mitotically dividing epithelial stem and progenitor cells in the embryonic lung show evidences of the asymmetric segregation and inheritance of Numb and asymmetric cell division similar to satellite muscle cells and neural stem cells (El-Hashash and Warburton, 2011, 2012). Lung epithelial stem and progenitor cells are also polarized, as evidenced by the apical localization of polarity proteins LGN, mInsc, NuMA and Par that regulate the orientation of spindle fibers in mitotically dividing epithelial cells of different organs in mammals (El-Hashash and Warburton, 2011, 2012; Berika *et al.*, 2014; Elshahawy *et al.*, 2016). In addition, most distal lung epithelial stem and progenitor cells have apically localized polarity proteins, with their mitotic spindles aligned perpendicular to the basement membrane (El-Hashash and Warburton, 2011, 2012; Berika *et al.*, 2014; Elshahawy *et al.*, 2016). Both the cell polarity exemplified by the polarized localization of Par, LGN, NuMA and mInsc proteins and perpendicular alignments of mitotic spindle fibers are closely related to and affect the asymmetric cell division in mammalian and *Drosophila* epithelial cell types and in other systems (Lechler and Fuchs, 2005; Zhu *et al.*, 2011; Kulukian and Fuchs 2013; Berika *et al.*, 2014; Yang *et al.*, 2015; di Pietro *et al.*, 2016).

Characterizing the mechanisms and functions of the cell fate determinant Numb in the lung is critical for better understanding of lung stem/progenitor cell behavior, morphogenesis, repair and regeneration. Numb is uniformly expressed during the interphase period of the cell cycle, and then it becomes localized to one cell of the two daughters during the asymmetric cell division. High Numb activity levels can suppress the cell differentiation, by inhibiting the activity of Notch signaling that acts to maintain stem the cell identity in different systems (Betschinger and Knoblich, 2004; Alexson *et al.*, 2006, Berika *et al.*, 2014; Elshahawy *et al.*, 2016). Cells that have low Numb activity can retain a high activity of Notch signaling, and therefore acquire a stem cell fate (Frise *et al.*, 1996; Guo *et al.*, 1996; Juven-Gershon *et al.*, 1998; Yan *et al.*, 2008; Berika *et al.*, 2014). In the embryonic lung, Numb is asymmetrically localized at the apical side of distal epithelial stem and progenitor cells and can play a major role in the determination of the mode of cell division: asymmetric vs. symmetric cell division. In addition, a correlation between the asymmetrical cell division and inheritance of Numb by one daughter cell was reported in distal lung epithelial stem and progenitor cells (El-Hashash and Warburton, 2011, 2012; Berika *et al.*, 2014; Elshahawy *et al.*, 2016) Furthermore, Numb knockdown in murine lung epithelial cells 15 (MLE15) results in high expression levels of stem cell markers Sox9/Id2, which further supports the function of Numb in lung epithelial cells (El-Hashash and Warburton, 2011, 2012).

The asymmetric mode of cell division plays a major role in the maintenance of the balance between cell self-renewal/proliferation and differentiation in various organs (Lechler and Fuchs, 2005; Powell *et al.*, 2010; Berika *et al.*, 2014; Elshahawy *et al.*, 2016). This balance is also important for normal regeneration of tissues in different organs, including lung. Thus, lack of balance between tissue self-renewal and differentiation of lung can result in diseases or disorders like bronchopulmonary dysplasia (BPD) and congenital lung hypoplasia, wherein a remarkable deficiency of stem and progenitor cells probably occurs, and these are general characteristics of prematurity and/or lung injury in humans. These lung diseases, disorders or abnormalities are common public health problems in newborn infants and a significant cause of death during infancy in humans. Therefore, adequate balance between cell proliferation/self-renewal and differentiation of stem/progenitor cells in the lung that is mediated by asymmetric cell division is highly required for the maintenance of normal lung growth and repair. The similarity in asymmetric stem/progenitor cell division between the lung and other organs may lead to identifying novel solutions for restoration of normal lung development and morphogenesis. However, this still need intensive studies on the mechanisms and regulatory factors of the asymmetric stem cell division in the lung.

Identification of novel signaling pathways and molecular mechanisms regulating the asymmetrical mode of cell division in lung epithelial stem cells, which is the mode controlling the balance between stem cell self-renewal and differentiation, will help defining new targets for the treatment of lethal lung abnormalities or diseases in infants and children and for lung repair after injury. The discovery of more molecular mechanisms that maintain the balance between proliferation and differentiation of lung stem and progenitor cells will play an important role in developing techniques for tackling the ability of stem cell to repair a diseased or injured lung.

The maturity of the fetal lung surfactant system is one of the two major steps to prepare the lung for air breathing after birth. During the last 8 weeks of human embryogenesis, fetal lung glycogen is broken down and converted into surfactant phospholipids. The most important of type of these surfactant phospholipids is disaturated phosphatidylcholine (DSPC). The process of lung maturation is tightly controlled by corticosteroids. This is evidenced by the research findings that null mutation of the glucocorticoid receptors and of corticotrophin can lead to the release of certain hormone(s) that can block the lung maturation in mice. Interestingly, human mutations in various components of the surfactant system have been reported, such as surfactant protein B (SP-B), which adversely affect the stability of pulmonary surfactant, and hence the ability to maintain lung inflation.

Remarkably, the transition to air breathing takes place rapidly in the mature neonatal lung of human infants. Thus, a significant increase in the level of catecholamine functions to switch off chloride secretion and induce sodium/potassium ATPase immediately following severance of the umblicial circulation (Olver and Strang, 1974, Olver *et al.*, 1986). This can act to reverse the production of tracheal fluid, leading to the rapid absorption of the tracheal fluid into the lung interestiitum and then into the lymphatic and pulmonary capillary circulation. Indeed, null mutation of Na/K ATPase in mice can lead to a failure in the absorption of fetal lung liquid which can result in a significant respiratory distress and even neonatal lethality (Hummler *et al.*, 1996).

References

Abe S, Lauby G, Boyer C, Rennard SI, Sharp JG. (2003). Transplanted BM and BM side population cells contribute progeny to the lung and liver in irradiated mice. *Cytotherapy* 5(6):523–533.

Adak S, Mukherjee S, Sen D. (2017). Mesenchymal stem cell as a potential therapeutic for inflammatory bowel disease — myth or reality? *Curr Stem Cell Res Ther*. doi: 10.2174/1574888X12666170914113633.

Adamson JYR. (1997). Development of lung structure. *The Lung*. Scientific Foundations, 2nd Edition. Lippincott-Raven Publishers, Philadelphia, p. 994.

Alberti C, Cochella L. (2017). A framework for understanding the roles of miRNAs in animal development. *Development* 144:2548–2559.

Alexson TO, Hitoshi S, Coles BL, Bernstein A, van der Kooy D. (2006). Notch signaling is required to maintain all neural stem cell populations — irrespective of spatial or temporal niche. *Dev Neurosci* 28(1–2):34–48.

Alphonse RS, Rajabali S, Thebaud B. (2012). Lung injury in preterm neonates: The role and therapeutic potential of stem cells Antioxid. *Redox Signal* 17(7):1013–1040.

An SS, Bai TR, Bates JH, Black JL, Brown RH, Brusasco V, Chitano P, Deng L, Dowell M, *et al.* (2007). Airway smooth muscle dynamics: A common pathway of airway obstruction in asthma. *Eur Respir J* 29:834–860.

Anselmo MA, Dalvin S, Prodhan P, Komatsuzaki K, Aidlen JT, Schnitzer JJ, Wu JY, Bernard KT. (2003). Slit and robo: Expression patterns in lung development. *Gene Expr Patterns* 3:13–19.

Antunes MA, Laffey JG, Pelosi P, Rocco PR. (2014). Mesenchymal stem cell trials for pulmonary diseases. *J Cell Biochem* 115:1023–1032.

Anversa P, Perrella MA, Kourembanas S, Choi AMK, Loscalzo J. (2012). Regenerative pulmonary medicine: Potential and promise, pitfalls and challenges. *Eur J Clin Invest* 42(8):900–913.

Araujo IM, Abreu SC, Maron-Gutierrez T, Cruz F, Fujisaki L, Carreira Jr., H, Ornellas F, Ornellas D, Vieira-de-Abreu A, Castro-Faria-Neto HC, Muxfeldt Ab'Saber A, Teodoro WR, Diaz BL, Peres Dacosta C, Capelozzi VL, Pelosi P, Morales MM, Rocco PR. (2010). Bone marrow-derived mononuclear cell therapy in experimental pulmonary and extrapulmonary acute lung injury. *Crit Care Med* 38(8):1733–1741.

Arora R, Metzger RJ, Papaioannou VE. (2012). Multiple roles and interactions of Tbx4 and Tbx5 in development of the respiratory system. *PLoS Genet* 8, e1002866.

Aslam R, Baveja OD, Liang A, Fernandez-Gonzalez C, Lee SA, *et al.* (2009). Bone marrow stromal cells attenuate lung injury in a murine model of neonatal chronic lung disease. *Am J Respir Crit Care Med* 180(11):1122–1130.

Asselin-Labat ML, Filby CE. (2012). Adult lung stem cells and their contribution to lung tumourigenesis. *Open Biology* 2(8):120094.

Aubin J, Lemieux M, Tremblay M, Berard J, Jeannotte L. (1997). Early postnatal lethality in Hoxa-5 mutant mice is attributable to respiratory tract defects. *Dev Biol* 192:432–445.

Awonusonu F, Srinivasan S, Strange J, Al Jumaily W, Bruce MC. (1999). Developmental shift in the relative percentages of lung fibroblast subsets: Role of apoptosis postseptation. *Am J Physiol* 277:L848–L859.

Baker CD, Balasubramaniam V, Mourani PM, Sontag MK, Black CP, Ryan SL. (2012). Cord blood angiogenic progenitor cells are decreased in bronchopulmonary dysplasia. *Eur Respir J* 40(6):1516–1522.

Balasubramaniam V, Mervis CF, Maxey AM, Markham NE, Abman SH. (2007). Hyperoxia reduces bone marrow, circulating, and lung endothelial progenitor cells in the developing lung: Implications for the pathogenesis of bronchopulmonary dysplasia. *Am J Physiol Lung Cell Mol Physiol* 292(5):L1073–L1084.

Balduino A, Mello-Coelho V, Wang Z, Taichman RS, Krebsbach PH, Weeraratna AT, Becker KG, de Mello W, Taub DD, Borojevic R. (2012). Molecular signature and *in vivo* behavior of bone marrow endosteal and subendosteal stromal cell populations and their relevance to hematopoiesis. *Exp Cell Res* 318(19):2427–2437.

Ballard PL, Ning Y, Polk D, Ikegami M, Jobe AH. (1997). Glucocorticoid regulation of surfactant components in immature lambs. *Am J Physiol Lung Cell Mol Physiol* 273(5 Pt 1):L1048–L1057.

Bansal G, Wong CM, Liu L, Suzuki YJ. (2012). Oxidant signaling for inter-leukin-13 gene expression in lung smooth muscle cells. *Free Radic Biol Med* 52(9):1552–1559.

Baraldi E, Filippone M. (2007). Chronic lung disease after premature birth. *N Engl J Med* 357(19):1946–1955.

Barkauskas CE, Cronce MJ, Rackley CR, Bowie EJ, Keene DR, Stripp BR, Randell SH, Noble PW, Hogan BL. (2013). Type 2 alveolar cells are stem cells in adult lung. *J Clin Invest* 123(7):3025–3036.

Bates JH, Lauzon AM. (2007). Parenchymal tethering, airway wall stiff-ness, and the dynamics of bronchoconstriction. *J Appl Physiol* 102:1912–1920.

Baveja R, Christou H. (2006). Pharmacological strategies in the preven-tion and management of bronchopulmonary dysplasia Semin. *Perinatol* 30(4):209–218.

Bellusci S, Grindley J, Emoto H, Itoh N, Hogan BL. (1997). Fibroblast growth factor 10 (FGF10) and branching morphogenesis in the embry-onic mouse lung. *Development* 124(23):4867–4878.

Bellusci S, Henderson R, Winnier G, Oikawa T, Hogan BL. (1996). Evidence from normal expression and targeted misexpression that bone morphogenetic protein (Bmp-4) plays a role in mouse embryonic lung morphogenesis. *Development* 122(6):1693–1702.

Bentley JK, Popova AP, Bozyk PD, Linn MJ, Baek AE, Lei J, Goldsmith AM, Hershenson MB. (2010). Ovalbumin sensitization and challenge increases the number of lung cells possessing a mesenchymal stromal cell phenotype. *Respir Res* 11:127.

Bentzinger CF, Wang YX, Rudnicki MA. (2012). Building muscle: Molecular regulation of myogenesis. *Cold Spring Harb Perspect Biol* 4(2).

Berger MJ, Adams SD, Tigges BM, Sprague SL, Wang XJ, Collins DP, McKenna DH. (2006). Differentiation of umbilical cord blood-derived multilineage progenitor cells into respiratory epithelial cells. *Cytotherapy* 8(5):480–487.

Berika M, Elgayyar M, El-Hashash A. (2014). Asymmetric cell divisions of stem cells in the lung and other systems. *Front Cell Dev Biol* (*Stem Cell Treatments*) 2:33.

Berkes CA, Tapscott SJ. (2005). MyoD and the transcriptional control of myogenesis. *Semin Cell Dev Biol* 16:585–595.

Bertram CD, Gaver DP. (2005). Bio-fluid mechanics of the pulmonary system. *Ann Biomed Eng* 33:1681–1688.

Betschinger J, Knoblich JA. (2004). Dare to be different: Asymmetric cell division in Drosophila, C. elegans and vertebrates. *Curr Biol* 14(16): R674–R685.

Bittmann I, Dose T, Baretton GB, Muller C, Schwaiblmair M, Kur F, Lohrs U. (2001). Cellular chimerism of the lung after transplantation. An interphase cytogenetic study. *Am J Clin Pathol* 115(4): 525–533.

Bokka KK, Jesudason EC, Lozoya OA, Guilak F, Warburton D, Lubkin SR. (2015). Morphogenetic implications of peristalsis-driven fluid flow in the embryonic lung. *PLoS One.*

Borger V, Bremer M, Ferrer-Tur R, Gockeln L, Stambouli O, Becic A, Giebel B. (2017). Mesenchymal stem/stromal cell-derived extracellular vesicles and their potential as novel immunomodulatory therapeutic agents. *International Journal of Molecular Sciences* 18(7).

Borthwick DW, Shahbazian M, Krantz QT, Dorin JR, Randell SH. (2001). Evidence for stem-cell niches in the tracheal epithelium. *Am J Respir Cell Mol Biol* 24(6):662–670.

Bostrom H, Gritli-Linde A, Betsholtz C. (2002). PDGF-A/PDGF alpha receptor signaling is required for lung growth and the formation of alveoli but not for early lung branching morphogenesis. *Dev Dyn* 223:155–162.

Bostrom H, Willetts K, Pekny M, Leveen P, Lindahl P, Hedstrand H, Pekna M, Hellstrom M, Gebre-Medhin S, Schalling M, Nilsson M, Kurland S, Tornell J, Heath JK, Betsholtz C. (1996). PDGF-A signaling is a critical event in lung alveolar myofibroblast development and alveogenesis. *Cell* 85:863–873.

Branchfield K, Li R, Lungova V, Verheyden JM, McCulley D, Sun X. (2016). A three-dimensional study of alveologenesis in mouse lung. *Dev Biol* 409:429–441.

Brennan J, Lu CC, Norris DP, Rodriguez TA, Beddington RS, Robertson EJ. (2001). Nodal signalling in the epiblast patterns the early mouse embryo. *Nature* 411:965–969.

Brody SL, Yan XH, Wuerffel MK, Song SK, Shapiro SD. (2000). Ciliogenesis and leftright axis defects in forkhead factor HFH-4-null mice. *Am J Respir Cell Mol Biol* 23:45–51.

Brown EM, Gamba G, Riccardi D, Lombardi M, Butters R, Kifor O, Sun A, Hediger MA, Lytton J, Hebert SC. (1993). Cloning and characterization of an extracellular Ca(2+)-sensing receptor from bovine parathyroid. *Nature* 366(6455):575–580.

Bryson-Richardson RJ, Currie PD. (2008). The genetics of vertebrate myogenesis. *Nat Rev Genet* 9(8):632–646.

Buckley S, Driscoll B, Anderson KD, Warburton D. (1997). Cell cycle in alveolar epithelial type II cells: Integration of Matrigel and KGF. *Am J Physiol* 273(3 Pt 1):L572–L580.

Buckley S, Barsky L, Weinberg K, Warburton D. (2005). *In vivo* inosine protects alveolar epithelial type 2 cells against hyperoxia-induced DNA damage through MAP kinase signaling. *Am J Physiol Lung Cell Mol Physiol* 288(3):L569–L575.

Burchell JT, Strickland DH, Stumbles PA. (2010). The role of dendritic cells and regulatory T cells in the regulation of allergic asthma. *Pharmacol Ther* 125(1):1–10.

Camargo FD, Chambers SM, Goodell MA. (2004). Stem cell plasticity: From transdifferentiation to macrophage fusion. *Cell Prolif* 37(1): 55–65.

Carthy JM, Garmaroudi FS, Luo Z, McManus BM. (2011). Wnt3a induces myofibroblast differentiation by upregulating TGF-β signaling through SMAD2 in a β-catenin-dependent manner. *PLoS One* 6(5):e19809.

Carraro G, El-Hashash A, Guidolin D, *et al.* (2009). miR-17 family of microRNAs controls FGF10-mediated embryonic lung epithelial branching morphogenesis through MAPK14 and STAT3 regulation of E-Cadherin distribution. *Dev Biol* 333(2):238–250.

Cavalcante FS, Ito S, Brewer K, Sakai H, Alencar AM, Almeida MP, Andrade Jr., JS, Majumdar A, Ingenito EP, Suki B. (2005). Mechanical interactions between collagen and proteoglycans: Implications for the stability of lung tissue. *J Appl Physiol* 98:672–679.

Cayouette M, Raff M. (2002). Asymmetric segregation of Numb: A mechanism for neural specification from Drosophila to mammals. *Nat Neurosci* 5(12):1265–1269.

Chang YS, Oh W, Choi SJ, Sung DK, Kim SY, Choi EY, Kang S, Jin HJ, Yang YS, Park WS. (2009). Human umbilical cord blood-derived mesenchymal stem cells attenuate hyperoxiainduced lung injury in neonatal rats. *Cell Transplant* 18(8):869–886.

Chapman HA, Li X, Alexander JP, Brumwell A, Lorizio W, Tan K, Sonnenberg A, Wei Y, Vu TH. (2011). Integrin alpha6beta4 identifies an adult distal lung epithelial population with regenerative potential in mice. *J Clin Invest* 121:2855–2862.

Chen L, Zosky GR. (2017). Lung development. *Photochem Photobiol Sci* 16:339–346.

Chen F, Desai TJ, Qian J, Niederreither K, Lü J, Cardoso WV. (2007). Inhibition of Tgf beta signaling by endogenous retinoic acid is essential for primary lung bud induction. *Development* 134(16):2969–2979.

Chen Y, Cao J, Qian F, Shao K, Niederreither Cardoso WV. (2010). A retinoic acid-dependent network in the foregut controls formation of the mouse lung primordium. *J Clin Invest* 120:2040–2048.

Chen Y, Schier AF. (2001). The zebrafish nodal signal squint functions as a morphogen. *Nature* 411:607–610.

Chiu WT, Charney Le R, Blitz IL, Fish MB, Li Y, Biesinger J, Xie X, Cho KW. (2014). Genome-wide view of tgfbeta/foxh1 regulation of the early mesendoderm program. *Development* 141:4537–4547.

Chokas AL, Trivedi CM, Lu MM, Tucker PW, Li S, Epstein JA, Morrisey EE. (2010). Foxp1/2/4-NuRD interactions regulate gene expression and epithelial injury response in the lung via regulation of interleukin-6. *J Biol Chem* 285:13304–13313.

Chow CW, Herrera Abreu MT, Suzuki T, Downey GP. (2003). Oxidative stress and acute lung injury. *Am J Respir Cell Mol Biol* 29(4):427–431.

Chuang PT, Kawcak T, McMahon AP. (2003). Feedback control of mammalian hedgehog signaling by the hedgehog-binding protein, Hip1, modulates Fgf signaling during branching morphogenesis of the lung. *Genes Dev* 17(3):342–347.

Chytil F. (1996). Retinoids in lung development. *FASEB J* 10:986–992.

Cohen JC, Larson JE, Killeen E, Love D, Takemaru K. (2008). CFTR and Wnt/beta-catenin signaling in lung development. *BMC Dev Biol* 8:70.

Coleman N, Phithakwatchara A, Shaaban S, Keswani B, Kline-Fath P, Kingma B, Haberman F, Lim Y. (2015). Fetal lung growth represented by longitudinal changes in MRI-derived fetal lung volume parameters predicts survival in isolated left-sided congenital diaphragmatic hernia. *Prenat Diagn* 35:160–166.

Colmenero J, Sancho-Bru P. (2017). Mesenchymal stromal cells for immunomodulatory cell therapy in liver transplantation: One step at a time. *J Hepatol* 67(1):7–9.

Colvin JS, Feldman B, Nadeau JH, Goldfarb M, Ornitz DM. (1999). Genomic organization and embryonic expression of the mouse fibroblast growth factor 9 gene. *Dev Dyn* 216:72–88.

Corcione A, Benvenuto F, Ferretti E, Giunti D, Cappiello V, Cazzanti F, Risso M, Gualandi F, Mancardi GL, Pistoia V, Uccelli A. (2006). Human mesenchymal stem cells modulate B-cell functions. *Blood* 107(1):367–372.

Daniely Y, Liao G, Dixon D, Linnoila RI, Lori A, Randell SH, Oren M, Jetten AM. (2004). Critical role of p63 in the development of a normal esophageal and tracheobronchial epithelium. *Am J Physiol Cell Physiol* 287(1):C171–C181.

Das BC, Thapa P, Karki R, Das S, Mahapatra S, Liu TC, *et al.* (2014). Retinoic acid signaling pathways in development and diseases. *Bioorg Med Chem* 22:673–683.

Davie NJ, Crossno Jr., JT, Frid MG, Hofmeister SE, Reeves JT, Hyde DM, Carpenter TC, Brunetti JA, McNiece IK, Stenmark KR. (2004). Hypoxia-induced pulmonary artery adventitial remodeling and neovascularization: Contribution of progenitor cells. *Am J Physiol Lung Cell Mol Physiol* 286(4):L668–L678.

De Langhe SP, Carraro G, Warburton D, Hajihosseini MK, Bellusci S. (2006). Levels of mesenchymal FGFR2 signaling modulate smooth muscle progenitor cell commitment in the lung. *Dev Biol* 299(1):52–62.

De Langhe SP, Carraro G, Tefft D, Li C, Xu X, Chai Y, Minoo P, Hajihosseini MK, Drouin J, Kaartinen V, Bellusci S. (2008). Formation and differentiation of multiple mesenchymal lineages during lung development is regulated by beta-catenin signaling. *PLoS One* 3(1):e1516.

De Moerlooze L, Spencer-Dene B, Revest J, Hajihosseini M, Rosewell I, Dickson C. (2000). An important role for the IIIb isoform of fibroblast growth factor receptor 2 (FGFR2) in mesenchymal-epithelial signalling during mouse organogenesis. *Development* 127:483–492.

del Moral PM, De Langhe SP, Sala FG, *et al.* (2006). Differential role of FGF9 on epithelium and mesenchyme in mouse embryonic lung. *Dev Biol* 293(1):77–89.

Desai TJ, Chen F, Lü J, Qian J, Niederreither K, Dollé P, Chambon P, Cardoso WV. (2006). Distinct roles for retinoic acid receptors alpha and beta in early lung morphogenesis. *Dev Biol* 291:12–24.

di Pietro F, Echard A, Morin X. (2016). Regulation of mitotic spindle orientation: An integrated view. *EMBO Reports* 17(8):1106–1130.

Domyan ET, Ferretti E, Throckmorton K, Mishina Y, Nicolis SK, Sun X. (2011). Signaling through BMP receptors promotes respiratory identity in the foregut via repression of Sox2. *Development* 138(5):971–981.

Donaldson SH, Bennett WD, Zeman KL, Knowles MR, Tarran R, Boucher RC. (2006). Mucus clearance and lung function in cystic fibrosis with hypertonic saline. *N Engl J Med* 354:241–250.

Driscoll B, Buckley S, Bui KC, Anderson KD, Warburton D. (2000). Telomerase in alveolar epithelial development and repair. *Am J Physiol Lung Cell Mol Physiol* 279(6):L1191–L1198.

Drubin DG, Nelson WJ. (1996). Origins of cell polarity. *Cell* 84:335–344.

Drummond ML, Prehoda KE. (2016). Molecular Control of atypical protein kinase C: Tipping the balance between self-renewal and differentiation. *J Mol Biol* 428(7):1455–1464.

Easley CA, Simerly CR, Schatten G. (2014). Gamete derivation from embryonic stem cells, induced pluripotent stem cells or somatic cell nuclear transfer-derived embryonic stem cells: State of the art. *Reprod Fertil Dev* 27(1):89–92.

Eblaghie MC, Reedy M, Oliver T, Mishina Y, Hogan BL. (2006). Evidence that autocrine signaling through Bmpr1a regulates the proliferation, survival and morphogenetic behavior of distal lung epithelial cells. *Dev Biol* 291(1):67–82.

El Agha E, Herold S, Al Alam D, *et al.* (2014). Fgf10-positive cells represent a progenitor cell population during lung development and postnatally. *Development* 141(2):296–306.

El-Hashash AHK. (2013). Lung stem/progenitor cells: Regulatory mechanisms of behavior, development and regeneration. *Anatom Physiol* 3:119.

El-Hashash AHK. (2014). New insights into the regulation and functional significance of Numb in lung stem cells during organogenesis. *Austin J Anatomy* 1(4):2.

El-Hashash AHK. (2018). Diversity of lung stem and progenitor cell types. *Lung Stem Cell Behavior.* First Edition. Springer Science Publisher, Heidelberg, Germany, p. 105.

El-Hashash AHK, Al Alam D, Turcatel G, Bellusci S, Warburton D. (2011). Eyes absent 1 (Eya1b) is a critical coordinator of epithelial, mesenchymal and vascular morphogenesis in the mammalian lung. *Dev Biol* 350(1): 112–126.

El-Hashash AHK, Al Alam D, Turcatel G, *et al.* (2011c). Six1 transcription factor is critical for coordination of epithelial, mesenchymal and vascular morphogenesis in the mammalian lung. *Dev Biol* 353(2):242–258.

El-Hashash AHK, Warburton D. (2011). Cell polarity and spindle orientation in the distal epithelium of embryonic lung. *Dev Dyn* 240(2):441–445.

El-Hashash AHK, Warburton D. (2012). Numb expression and asymmetric versus symmetric cell division in distal embryonic lung epithelium. *J Histochem Cytochem* 60(9):675–682.

Ellis T, Gambardella L, Horcher M, Tschanz S, Capol J, Bertram P, Jochum W, Barrandon Y, Busslinger M. (2001). The transcriptional repressor CDP (Cutl1) is essential for epithelial cell differentiation of the lung and the hair follicle. *Genes Dev* 15(17):2307–2319.

Elshahawy S, Ibrahim A, Berika M, El-Hashash AH. (2016). Behavior and asymmetric cell divisions of stem cells. *Developmental and Stem Cell Biology in Health and Disease.* Vol.1, Bentham Science Publisher, USA, pp. 81–104.

Engelhardt JF. (2001). Stem cell niches in the mouse airway. *Am J Respir Cell Mol Biol* 24(6):649–652.

Engler AJ, Sen S, Sweeney HL, Discher DE. (2006). Matrix elasticity directs stem cell lineage specification. *Cell* 126(4):677–689.

Epperly MW, Guo H, Gretton JE, Greenberger JS. (2003). Bone marrow origin of myofibroblasts in irradiation pulmonary fibrosis. *Am J Respir Cell Mol Biol* 29(2):213–224.

Evans CM, Williams OW, Tuvim MJ, Nigam R, Mixides GP, Blackburn MR, DeMayo FJ, Burns AR, Smith C, Reynolds SD, *et al.* (2004). Mucin is produced by clara cells in the proximal airways of antigen-challenged mice. *Am J Respir Cell Mol Biol* 31:382–394.

Evans MJ, Cabral LJ, Stephens RJ, Freeman G. (1975). Transformation of alveolar type 2 cells to type 1 cells following exposure to NO_2. *Exp Mol Pathol* 22(1):142–150.

Featherstone NC, Jesudason EC, Connell MG, *et al.* (2005). Spontaneous propagating calcium waves underpin airway peristalsis in embryonic rat lung. *Am J Resp Cell and Mol Biol* 33(2):153–160.

Fewell JE, Hislop AA, Kitterman JA, Johnson P. (1983). Effect of tracheostomy on lung development in fetal lambs. *J Appl Physiol Respir Environ Exerc Physiol* 55(4):1103–1108.

Finney BA, del Moral PM, Wilkinson WJ, Cayzac S, Cole M, Warburton D, Kemp PJ, Riccardi D. (2008). Regulation of mouse lung development by the extracellular calcium-sensing receptor, CaR. *J Physiol* 586(Pt 24): 6007–6019.

Fischer A, Viebahn C, Blum, M. (2002). FGF8 acts as a right determinant during establishment of the left-right axis in the rabbit. *Curr Biol* 12(21):1807–1816.

Folkesson HG, Norlin A, Wang Y, Abedinpour P, Matthay MA. (2000). Dexamethasone and thyroid hormone pretreatment upregulate alveolar epithelial fluid clearance in adult rats. *J Appl Physiol* 88:416–424.

Forgacs G, Foty RA, Shafrir Y, Steinberg MS. (1998). Viscoelastic properties of living embryonic tissues: A quantitative study. *Biophys J* 74:2227–2234.

Foty RA, Forgacs G, Pfleger CM, Steinberg MS. (1994). Liquid properties of embryonic tissues: Measurement of interfacial tensions. *Phys Rev Lett* 72:2298–2301.

Fredberg JJ, Kamm RD. (2006). Stress transmission in the lung: Pathways from organ to molecule. *Annu Rev Physiol* 68:507–541.

Frise E, Knoblich J, Younger-Shepherd S, Jan LY, Jan YN. (1996). The Drosophila Numb protein inhibits signaling of the Notch receptor during cell-cell interaction in sensory organ lineage. *Proc Natl Acad Sci USA* 93(21):11925–11932.

Fust A, LeBellego F, Iozzo RV, Roughley PJ, Ludwig MS. (2005). Alterations in lung mechanics in decorin-deficient mice. *Am J Physiol Lung Cell Mol Physiol* 288:L159–L166.

Gao F, Chiu SM, Motan DA, Zhang Z, Chen L, Ji HL, Tse HF, Fu QL, Lian Q. (2016). Mesenchymal stem cells and immunomodulation: Current status and future prospects. *Cell Death Dis* 7:e2062.

Gao X, Vockley CM, Pauli F, *et al.* (2013). Evidence for multiple roles for grainyheadlike 2 in the establishment and maintenance of human mucociliary airway epithelium. *Proc Natl Acad Sci USA* 110(23):9356–9361.

Gao Y, Raj JU. (2005). Parathyroid hormone-related protein-mediated responses in pulmonary arteries and veins of newborn lambs. *Am J Physiol Lung Cell Mol Physiol* 289(1):L60–L66.

Geng Y, Dong Y, Yu M, Zhang L, Yan X, Sun J, Qiao L, Geng H, Nakajima M, Furuichi T, Ikegawa S, Gao X, Chen YG, Jiang D, Ning W. (2011). Follistatin-like 1 (Fstl1) is a bone morphogenetic protein (BMP) 4 signaling antagonist in controlling mouse lung development. *Proc Natl Acad Sci USA* 108(17):7058–7063.

Giangreco A, Reynolds SD, Stripp BR. (2002). Terminal bronchioles harbor a unique airway stem cell population that localizes to the bronchoalveolar duct junction. *Am J Pathol* 161(1):173–182.

Giangreco A, Shen H, Reynolds SD, Stripp BR. (2004). Molecular phenotype of airway side population cells. *Am J Physiol Lung Cell Mol Physiol* 286(4):L624–L630.

Glennie S, Soeiro I, Dyson PJ, Lam EW, Dazzi F. (2005). Bone marrow mesenchymal stem cells induce division arrest anergy of activated T cells. *Blood* 105(7):2821–2827.

Gill SE, Pape MC, Khokha R, Watson AJ, Leco KJ. (2003). A null mutation for tissue inhibitor of metalloproteinases-3 (Timp-3) impairs murine bronchiole branching morphogenesis. *Dev Biol* 261:313–323.

Goldstein JD, Reid LM. (1980). Pulmonary hypoplasia resulting from phrenic nerve agenesis and diaphragmatic amyoplasia. *J Pediatr* 97(2):282–287.

Goldberger AL, West BJ. (1987). Fractals in physiology and medicine. *Yale J Biol Med* 60(5):421–435.

Goss AM, Tian Y, Tsukiyama T, Cohen ED, Zhou D, Lu MM, Yamaguchi TP, Morrisey EE. (2009). Wnt2/2b and beta-catenin signaling are necessary and sufficient to specify lung progenitors in the foregut. *Dev Cell* 17(2):290–298.

Goss AM, Tian Y, Cheng L, *et al.* (2011). Wnt2 signaling is necessary and sufficient to activate the airway smooth muscle program in the lung by regulating myocardin/Mrtf-B and Fgf10 expression. *Dev Biol* 356(2):541–552.

Green MD, Chen A, Nostro MC, D'Souza SL, Schaniel C, Lemischka IR, Gouon-Evans V, Keller G, Snoeck HW. (2011). Generation of anterior foregut endoderm from human embryonic and induced pluripotent stem cells. *Nat Biotechnol* 29(3):267–272.

Green MD, Huang SX, Snoeck HW. (2013). Stem cells of the respiratory system: From identification to differentiation into functional epithelium. *Bioessays* 35(3):261–270.

Grove JE, Lutzko C, Priller J, Henegariu O, Theise ND, Kohn DB, Krause DS. (2002). Marrow-derived cells as vehicles for delivery of gene therapy to pulmonary epithelium. *Am J Respir Cell Mol Biol* 27(6): 645–651.

Guo M, Jan LY, Jan YN. (1996). Control of daughter cell fates during asymmetric division: Interaction of Numb and Notch. *Neuron* 17(1):27–41.

Gupte VV, Ramasamy SK, Reddy R, *et al.* (2009). Overexpression of fibroblast growth factor-10 during both inflammatory and fibrotic phases attenuates bleomycin-induced pulmonary fibrosis in mice. *Am J Respir Crit Care Med* 180(5):424–436.

Guseh JS, Bores SA, Stanger BZ, Zhou Q, Anderson WJ, Melton DA, Rajagopal J. (2009). Notch signaling promotes airway mucous metaplasia and inhibits alveolar development. *Development* 136:1751–1759.

Hackett TL, Shaheen F, Johnson A, *et al.* (2008). Characterization of side population cells from human airway epithelium. *Stem Cells* 26(10): 2576–2585.

Harding R. (1997). *Fetal Breathing Movements.* 2nd edn., Crystal RG, West JB, Banes PJ, Weiber ER (editors). The Lung: Scientific Foundations, Philadelphia, Lippincott-Raven, 2093–2104.

Harding R, Hooper SB. (1996). Regulation of lung expansion and lung growth before birth. *J Appl Physiol* 81(1):209–224.

Harris-Johnson KS, Domyan ET, Vezina CM, Sun X. (2009). Beta-catenin promotes respiratory progenitor identity in mouse foregut. *Proc Natl Acad Sci USA* 106:16287–16292.

Hashimoto N, Jin H, Liu T, Chensue SW, Phan SH. (2004). Bone marrow-derived progenitor cells in pulmonary fibrosis. *J Clin Invest* 113(2): 243–252.

Hashimoto S, Nakano H, Singh G, Katyal S. (2002). Expression of Spred and Sprouty in developing rat lung. *Mech Dev* 119(Suppl 1): S303–S309.

Hashimshony T, Feder M, Levin M, Hall BK, Yanai I. (2015). Spatiotemporal transcriptomics reveals the evolutionary history of the endoderm germ layer. *Nature* 519:219–222.

Hastings RH. (2004). Parathyroid hormone-related protein and lung biology. *Resp Physiol Neurobiol* 142(2–3):95–113.

Haydar TF, Ang Jr., E, Rakic, P. (2003). Mitotic spindle rotation and mode of cell division in the developing telencephalon. *Proc Natl Acad Sci USA* 100:2890–2895.

Hegab AE, Ha VL, Gilbert JL, *et al.* (2011). Novel stem/progenitor cell population from murine tracheal submucosal gland ducts with multipotent regenerative potential. *Stem Cells* 29(8):1283–1293.

Herriges MJ, Swarr DT, Morley MP, Rathi KS, Peng T, Stewart KM, Morrisey EE. (2014). Long noncoding RNAs are spatially correlated with transcription factors and regulate lung development. *Genes* 1363–1379.

Herzog EL, Chai L, Krause DS. (2003). Plasticity of marrow-derived stem cells. *Blood* 102(10):3483–3493.

Hogan BLM, Barkauskas CE, Chapman HA, *et al.* (2014). Repair and regeneration of the respiratory system: Complexity, plasticity, and mechanisms of lung stem cell function. *Cell Stem Cell* 15(2): 123–138.

Hodges RJ, Lim R, Jenkin G, Wallace EM. (2012). Amnion epithelial cells as a candidate therapy for acute and chronic lung injury. *Stem Cells Int* 709–763.

Holgate ST, Wenzel S, Postma DS, Weiss ST, Renz H, Sly PD. (2015). Asthma. *Nat Rev Dis Primers* 1:15025.

Hong KU, Reynolds SD, Giangreco A, Hurley CM, Stripp BR. (2001). Clara cell secretory protein-expressing cells of the airway neuroepithelial body microenvironment include a label-retaining subset and are critical for epithelial renewal after progenitor cell depletion. *Am J Respir Cell Mol Biol* 24(6):671–681.

Hong KU, Reynolds SD, Watkins S, Fuchs E, Stripp BR. (2004). Basal cells are a multipotent progenitor capable of renewing the bronchial epithelium. *Am J Pathol* 164(2):577–588.

Hooper SB, Wallace MJ. (2006). Role of the physicochemical environment in lung development. *Clin Exp Pharmacol Physiol* 33(3):273–279.

Hori S, Nomura T, Sakaguchi S. (2003). Control of regulatory T cell development by the transcription factor Foxp3. *Science* 299(5609): 1057–1061.

Hsu YC, Osinski J, Campbell CE, Litwack ED, Wang D, Liu S, Bachurski CJ, Gronostajski RM. (2011). Mesenchymal nuclear factor I B regulates cell proliferation and epithelial differentiation during lung maturation. *Dev Biol* 354(2):242–252.

Huang Y, Shen XJ, Zou Q, Wang SP, Tang SM, Zhang GZ. (2011). Biological functions of microRNAs: A review. *J Physiol Biochem* 67(1):129–139. doi:10.1007/s13105-010-0050-6.

Huang Z, Wang Y, Nayak PS, Dammann CE, Sanchez-Esteban J. (2012). Stretch-induced fetal type II cell differentiation is mediated via ErbB1-ErbB4 interactions. *J Biol Chem* 287(22):18091–18102.

Hummler E, Barker P, Gatzy J, Beermann F, Verdumo C, Schmidt A, Boucher R, Rossier BC. (1996). Early death due to defective neonatal lung liquid clearance in alpha-ENaC-deficient mice. *Nat Genet* 12(3): 325–328.

Huttmann A, Li CL, Duhrsen U. (2003). Bone marrow-derived stem cells and "plasticity". *Ann Hematol* 82(10):599–604.

Huttner WB, Kosodo Y. (2005). Symmetric versus asymmetric cell division during neurogenesis in the developing vertebrate central nervous system. *Curr Opin Cell Biol* 17(6):648–657.

Husain AN, Hessel RG. (1993). Neonatal pulmonary hypoplasia: An autopsy study of 25 cases. *Pediatr Pathol* 13(4):475–484.

Ibrahim A, El-Hashash AH. (2015). Lung stem cell behavior in development and regeneration. *Edorium J Stem Cell Res Ther* 1:1–13.

Inanlou MR, Kablar B. (2003). Abnormal development of the diaphragm in mdx: MyoD−/−(9th) embryos leads to pulmonary hypoplasia. *Int J Dev Biol* 47(5):363–371.

Ionescu L, Byrne RN, van Haaften T, Vadivel A, Alphonse RS, Rey-Parra GJ, Weissmann G, Hall A, Eaton F, Thebaud B. (2012). Stem cell conditioned medium improves acute lung injury in mice: *In vivo* evidence for stem cell paracrine action. *Am J Physiol Lung Cell Mol Physiol* 303(11): L967–L977.

Iosef C, Alastalo TP, Hou Y, Chen C, Adams ES, Lyu SC, Cornfield DN, Alvira CM. (2012). Inhibiting NF-κB in the developing lung disrupts angiogenesis and alveolarization. *Am J Physiol Lung Cell Mol Physiol* 302(10):L1023–L1036.

Irwin D, Helm K, Campbell N, *et al.* (2007). Neonatal lung side population cells demonstrate endothelial potential and are altered in response to hyperoxia-induced lung simplification. *Am J Physiol Lung Cell Mol Physiol* 293(4):L941–L951.

Ishizawa K, Kubo H, Yamada M, Kobayashi S, Numasaki M, Ueda S, Suzuki T, Sasaki H. (2004). Bone marrow-derived cells contribute to lung regeneration after elastase-induced pulmonary emphysema. *FEBS Lett* 556(1–3):249–252.

Ito S, Bartolak-Suki E, Shipley JM, Parameswaran H, Majumdar A, Suki B. (2006). Early emphysema in the tight skin and pallid mice: Roles of microfibrilassociated glycoproteins, collagen, and mechanical forces. *Am J Respir Cell Mol Biol* 34:688–694.

Ito T, Udaka N, Yazawa T, Okudela K, Hayashi H, Sudo T, Guillemot F, Kageyama R, Kitamura H. (2000). Basic helix-loop-helix transcription factors regulate the neuroendocrine differentiation of fetal mouse pulmonary epithelium. *Development* 127(18):3913–3921.

Jakab K, Damon B, Marga F, Doaga O, Mironov V, Kosztin I, Markwald R, Forgacs G (2008a). Relating cell and tissue mechanics: Implications and applications. *Dev Dyn* 237:2438–2449.

Jakab K, Norotte C, Damon B, Marga F, Neagu A, Besch-Williford CL, Kachurin A, Church KH, Park H, Mironov V, Markwald RR, Vunjak-Novakovic G, Forgacs G. (2008b). Tissue engineering by self-assembly of cells printed into topologically defined structures. *Tissue Eng* 14:413–421.

Jeffery PK. (2001). Remodeling in asthma and chronic obstructive lung disease. *Am J Respir Crit Care Med* 164(10 Pt 2):S28–S38.

Jesudason EC. (2006). Small lungs and suspect smooth muscle: Congenital diaphragmatic hernia and the smooth muscle hypothesis. *J Pediatr Surg* 41(2):431–435.

Jesudason EC. (2009). Airway smooth muscle: An architect of the lung? *Thorax* 64(6):541–545.

Jesudason EC, Connell MG, Fernig DG, Lloyd DA, Losty PD. (2000). Early lung malformations in congenital diaphragmatic hernia. *J Pediatr Surg* 35:124–127.

Jesudason EC, Smith NP, Connell MG, Spiller DG, White MR, Fernig DG, Losty PD. (2005). Developing rat lung has a sided pacemaker region for morphogenesis-related airway peristalsis. *Am J Respir Cell Mol Biol* 32:118–127.

Jesudason R, Black L, Majumdar A, Stone P, Suki B. (2007). Differential effects of static and cyclic stretching during elastase digestion on the mechanical properties of extracellular matrices. *J Appl Physiol* 103:803–811.

Jiang Y, Jahagirdar BN, Reinhardt RL, Schwartz RE, Keene CD, Ortiz-Gonzalez XR, Reyes M, Lenvik T, Lund T, Blackstad M, Du J, Aldrich S, Lisberg A, Low WC, Largaespada DA, Verfaillie CM. (2002). Pluripotency of mesenchymal stem cells derived from adult marrow. *Nature* 418(6893):41–49.

Jiang JX, Li L. (2009). Potential therapeutic application of adult stem cells in acute respiratory distress syndrome. *Chin J Traumatol* 12(4):228–233. [Edorium *Journal of Stem Cell Research and Therapy*, Vol. 1; 2015. Edorium *J Stem Cell Res Ther* 2015. 1:1–13], *www.edoriumjournals.com/ej/srt* Ibrahim *et al.* 10.

Jobe AJ. (1999). The new BPD: An arrest of lung development. *Pediatr Res* 46(6):641.

Juven-Gershon T, Shifman O, Unger T, Elkeles A, Haupt Y, Oren M. (1998). The Mdm2 oncoprotein interacts with the cell fate regulator Numb. *Mol Cell Biol* 18(7):3974–3978.

Kaartinen V, Voncken JW, Shuler C, Warburton D, Bu D, Heisterkamp N, Groffen J. (1995). Abnormal lung development and cleft palate in mice lacking TGF-beta 3 indicates defects of epithelial-mesenchymal interaction. *Nat Genet* 11:415–421.

Kasahara Y, Tuder RM, Taraseviciene-Stewart L, Le Cras TD, Abman S, Hirth PK, Waltenberger J, Voelkel NF. (2000). Inhibition of VEGF receptors causes lung cell apoptosis and emphysema. *J Clin Invest* 106:1311–1319.

Keating A. (2012). Mesenchymal stromal cells: New directions. *Cell Stem Cell* 10(6):709–716.

Kelly GM, Drysdale TA. (2015). Retinoic acid and the development of the endoderm. *J Dev Biol* 3:25–56.

Keijzer R, Liu J, Deimling J, Tibboel D, Post M. (2000). Dual-hit hypothesis explains pulmonary hypoplasia in the nitrofen model of congenital diaphragmatic hernia. *Am J Pathol* 156:1299–1306.

Khan FM, Sy S, Louie P, Smith M, Chernos J, Berka N, Sinclair GD, Lewis V, Russell JA, Storek J. (2010). Nasal epithelial cells of donor origin after allogeneic hematopoietic cell transplantation are generated at a faster rate in the first 3 months compared with later posttransplantation. *Biol Blood Marrow Transplant* 16(12):1658–1664.

Kim CF, Jackson EL, Woolfenden AE, *et al.* (2005). Identification of bronchioalveolar stem cells in normal lung and lung cancer. *Cell* 121(6):823–835.

Kim EJY, Korotkevich E, Hiiragi T. (2018). Coordination of cell polarity, mechanics and fate in tissue self-organization. *Trends Cell Biol* pii: S0962–S8924(18)30032-1.

Kim ES, Chang YS, Choi SJ, Kim JK, Yoo HS, Ahn SY, Sung DK, Kim SY, Park YR, Park WS. (2011). Intratracheal transplantation of human umbilical cord blood-derived mesenchymal stem cells attenuates Escherichia coli-induced acute lung injury in mice. *Respir Res* 12:108.

Kim N, Vu TH. (2006). Parabronchial smooth muscle cells and alveolar myofibroblasts in lung development. *Birth Defects Res C Embryo Today* 78(1):80–89.

Kim SJ, Habib O, Kim JS, Han HW, Koo SK, Kim JH. (2017). Generation of a Nrf2 homozygous knockout human embryonic stem cell line using CRISPR/Cas9. *Stem Cell Research* 19:46–48.

Kim TH, Shivdasani RA. (2016). Stomach development, stem cells and disease. *Development* 143(4):554–565.

Kimura S, Hara Y, Pineau T, Fernandez-Salguero P, Fox CH, Ward JM, Gonzalez FJ. (1996). The T/ebp null mouse: Thyroid-specific enhancer-binding protein is essential for the organogenesis of the thyroid, lung, ventral forebrain, and pituitary. *Genes Dev* 10:60–69.

Kimura S, Ward JM, Minoo P. (1999). Thyroid-specific enhancer-binding protein/thyroid transcription factor 1 is not required for the initial specification of the thyroid and lung primordia. *Biochimie* 81(4):321–327.

Kitagawa M, Takebe A, Ono Y, Imai T, Nakao K, Nishikawa SI, Era T. (2012). Phf14, a Novel regulator of mesenchyme growth via platelet-derived growth factor (PDGF) receptor-α. *J Biol Chem* 287(33): 27983–27996.

Kitamura K, Miura H, Miyagawa-Tomita S, Yanazawa M, Katoh-Fukui Y, Suzuki R, Ohuchi H, Suehiro A, Motegi Y, Nakahara Y, Kondo S, Yokoyama M. (1999). Mouse Pitx2 deficiency leads to anomalies of the ventral body wall, heart, extra- and periocular mesoderm and right pulmonary isomerism. *Development* 126(24):5749–5758.

Kitaoka H, Burri PH, Weibel ER. (1996). Development of the human fetal airway tree: Analysis of the numerical density of airway end tips. *Anat Rec* 244(2):207–213.

Kitterman JA. (1996). The effects of mechanical forces on fetal lung growth. *Clin Perinatol* 23(4):727–740.

Kleeberger W, Versmold A, Rothamel T, Glockner S, Bredt M, Haverich A, Lehmann U, Kreipe H. (2003). Increased chimerism of bronchial and alveolar epithelium in human lung allografts undergoing chronic injury. *Am J Pathol* 162(5):1487–1494.

Komatsu Y, Shibuya H, Takeda N, Ninomiya-Tsuji J, Yasui T, Miyado K, Sekimoto T, Ueno N, Matsumoto K, Yamada G. (2002). Targeted disruption of the Tab1 gene causes embryonic lethality and defects in cardiovascular and lung morphogenesis. *Mech Dev* 119:239–249.

Korbling M, Estrov Z. (2003). Adult stem cells for tissue repair — a new therapeutic concept? *N Engl J Med* 349(6):570–582.

Knoblich JA. (2001). Asymmetric cell division during animal development. *Nat Rev Mol Cell Biol* 2(1):11–20, 349(6):570–582.

Kosodo Y, Röper K, Haubensak W, Marzesco AM, Corbeil D, Huttner WB. (2004). Asymmetric distribution of the apical plasma membrane during neurogenic divisions of mammalian neuroepithelial cells. *EMBO J* 23(11):2314–2324.

Kotton DN, Fabian AJ, Mulligan RC. (2005). Failure of bone marrow to reconstitute lung epithelium. *Am J Respir Cell Mol Biol* 33(4):328–334.

Krause DS, Theise ND, Collector MI, Henegariu O, Hwang S, Gardner R, Neutzel S, Sharkis SJ. (2001). Multi-organ, multi-lineage engraftment by a single bone marrow-derived stem cell. *Cell* 105(3):369–377.

Kreidberg JA, Donovan MJ, Goldstein SL, Rennke H, Shepherd K, Jones RC, Jaenisch R. (1996). Alpha 3 beta 1 integrin has a crucial role in kidney and lung organogenesis. *Development* 122(11):3537–3547.

Kruithof-de Julio M, Alvarez MJ, Galli A, Chu J, Price SM, Califano A, Shen MM. (2011). Regulation of extra-embryonic endoderm stem cell differentiation by nodal and cripto signaling. *Development* 138: 3885–3895.

Ku J, El-Hashash AH. (2016). Molecular control of the mode of cell division: A view from mammalian lung epithelial stem cells. *Journal of Anatomy* 3:3–6.

Kulukian A, Fuchs E. (2013). Spindle orientation and epidermal morphogenesis. *Phil Trans R Soc B* 368:20130016.

Lechler T, Fuchs E. (2005). Asymmetric cell divisions promote stratification and differentiation of mammalian skin. *Nature* 437(7056):275–280.

Lee JW, Fang X, Krasnodembskaya A, Howard JP, Matthay MA. (2011). Concise review: Mesenchymal stem cells for acute lung injury: Role of paracrine soluble factors. *Stem Cells* 29(6):913–919.

Lee JW, Krasnodembskaya A, McKenna DH, Song Y, Abbott J, Matthay MA. (2013). Therapeutic effects of human mesenchymal stem cells in ex vivo human lungs injured with live bacteria. *Am J Respir Crit Care Med* 187(7):751–760.

Lee RH, Pulin AA, Seo MJ, Kota DJ, Ylostalo J, Larson BL, Semprun-Prieto L, Delafontaine P, Prockop DJ. (2009). Intravenous hMSCs improve myocardial infarction in mice because cells embolized in lung are activated to secrete the anti-inflammatory protein TSG-6. *Cell Stem Cell* 5(1):54–63.

Lefrancais E, Ortiz-Munoz G, Caudrillier A, Mallavia B, Liu F, Sayah DM, Thornton EE, Headley MB, David T, Coughlin SR, Krummel MF, Leavitt AD, Passegue E, Looney MR. (2017). The lung is a site of platelet biogenesis and a reservoir for haematopoietic progenitors. *Nature* 544(7648):105–109.

Lemaire P, Darras S, Caillol D, Kodjabachian L. (1998). A role for the vegetally expressed xenopus gene mix.1 in endoderm formation and in the restriction of mesoderm to the marginal zone. *Development* 125: 2371–2380.

Leung LY, Tian D, Brangwynne CP, Weitz DA, Tschumperlin DJ. (2007). A new microrheometric approach reveals individual and cooperative roles for TGFbeta1 and IL-1beta in fibroblast-mediated stiffening of collagen gels. *Faseb J* 21:2064–2073.

Li C, Xiao J, Hormi K, Borok Z, Minoo P. (2002). Wnt5a participates in distal lung morphogenesis. *Dev Biol* 248(1):68–81.

Li JW, Wu X. (2015). Mesenchymal stem cells ameliorate LPS-induced acute lung injury through KGF promoting alveolar fluid clearance of alveolar type II cells. *Eur Rev Med Pharmacol Sci* 19(13):2368–2378.

Li S, Weidenfeld J, Morrisey EE. (2004). Transcriptional and DNA binding activity of the Foxp1/2/4 family is modulated by heterotypic and homotypic protein interactions. *Mol Cell Biol* 24:809–822.

Li S, Xiang, M. (2011). Foxn4 influences alveologenesis during lung development. *Dev Dyn* 240(6):1512–1517.

Lia S, Morleya M, Lua M, Zhoua S, Stewarta K, Frenchf CA, Tuckerg HO, Fisherh SE, Morriseya EE. (2016). Foxp transcription factors suppress a non-pulmonary gene expression program to permit proper lung development. *Develop Biol* 416(2):338–346.

Lim L, Kalinichenko VV, Whitsett JA, Costa RH. (2002). Fusion of lung lobes and vessels in mouse embryos heterozygous for the forkhead box f1 targeted allele. *Am J Physiol Lung Cell Mol Physiol* 282(5):L1012–L1022.

Lindahl P, Karlsson L, Hellstrom M, Gebre-Medhin S, Willetts K, Heath JK, Betsholtz C. (1997). Alveogenesis failure in PDGF-A-deficient mice is coupled to lack of distal spreading of alveolar smooth muscle cell progenitors during lung development. *Development* 124:3943–3953.

Litingtung Y, Lei L, Westphal H, Chiang C. (1998). Sonic hedgehog is essential to foregut development. *Nat Genet* 20:58–61.

Little MH. (2011). Renal organogenesis: What can it tell us about renal repair and regeneration? *Organogenesis* 7(4):229–241.

Liu C, Glasser SW, Wan H, Whitsett JA. (2002). GATA-6 and thyroid transcription factor-1 directly interact and regulate surfactant protein-C gene expression. *J Biol Chem* 277(6):4519–4525.

Liu JP, Baker J, Perkins AS, Robertson EJ, Efstratiadis A. (1993). Mice carrying null mutations of the genes encoding insulin-like growth factor I (Igf-1) and type 1 IGF receptor (Igf1r). *Cell* 75:59–72.

Liu M, Post M. (2000). Invited review: Mechanochemical signal transduction in the fetal lung. *J Appl Physiol* 89(5):2078–2084.

Liu M, Zeng X, Wang J, Fu Z, Wang J, Liu M, Ren D, Yu B, Zheng L, Hu X, Shi W, Xu J. (2016). Immunomodulation by mesenchymal stem cells in treating human autoimmune disease-associated lung fibrosis. *Stem Cell Res Ther* 7(1):63.

Liu P, Wakamiya M, Shea MJ, Albrecht U, Behringer RR, Bradley A. (1999). Requirement for wnt3 in vertebrate axis formation. *Nat Genet* 22:361–365.

Liu W, Brown K, Legros S, Foley AC. (2012). Nodal mutant extraembryonic endoderm (xen) stem cells upregulate markers for the anterior visceral endoderm and impact the timing of cardiac differentiation in mouse embryoid bodies. *Biol Open* 1:208–219.

Liu Y, Hogan BL. (2002). Differential gene expression in the distal tip endoderm of the embryonic mouse lung. *Gene Expr Patterns* 2(3–4): 229–233.

Liu Y, Jiang H, Crawford HC, Hogan BL. (2003). Role for ETS domain transcription factors Pea3/Erm in mouse lung development. *Dev Biol* 261:10–24.

Livraghi-Butrico A, Grubb BR, Wilkinson KJ, *et al.* (2017). Contribution of mucus concentration and secreted mucins Muc5ac and Muc5b to the pathogenesis of muco-obstructive lung disease. *Mucosal Immunology* 10(2):395–407.

Londhe VA, Maisonet TM, Lopez B, Shin BC, Huynh J, Devaskar SU. (2013). Retinoic acid rescues alveolar hypoplasia in the calorie-restricted developing rat lung. *Am J Respir Cell Mol Biol* 48:179–187.

Longmire TA, Ikonomou L, Hawkins F, Christodoulou C, Cao YX, Jean JC, Kwok LW, Mou HM, Rajagopal J, Shen SS, Dowton AA, Serra M, Weiss DJ, Green MD, Snoeck HW, Ramirez MI, Kotton DN. (2012). Efficient derivation of purified lung and thyroid progenitors from embryonic stem cells. *Cell Stem Cell* 10(4):398–411.

Lordan JL, Bucchieri F, Richter A, Konstantinidis A, Holloway JW, Thornber M, Puddicombe SM, Buchanan D, Wilson SJ, Djukanovic R, Holgate ST, Davies DE. (2002). Cooperative effects of Th2 cytokines and allergen on normal and asthmatic bronchial epithelial cells. *J Immunol* 169:407–414.

Love D, Li FQ, Burke MC, Cyge B, Ohmitsu M, Cabello J, Larson JE, Brody SL, Cohen JC, Takemaru K. (2010). Altered lung morphogenesis, epithelial cell differentiation and mechanics in mice deficient in the Wnt/β-catenin antagonist Chibby. *PLoS One* 5(10):e13600.

Lowe LA, Yamada S, Kuehn MR. (2001). Genetic dissection of nodal function in patterning the mouse embryo. *Development* 128(10): 1831–1843.

Lu MM, Li S, Yang H, Morrisey EE. (2002). Foxp4: A novel member of the Foxp subfamily of winged-helix genes co-expressed with Foxp1 and Foxp2 in pulmonary and gut tissues. *Mech Dev* 119(Suppl 1): S197–S202.

Lu MM, Yang H, Zhang L, Shu W, Blair DG, *et al.* (2001). The bone morphogenic protein antagonist gremlin regulates proximal-distal patterning of the lung. *Dev Dyn* 222:667–680.

Lu Y, Okubo T, Rawlins E, Hogan BL. (2008). Epithelial progenitor cells of the embryonic lung and the role of microRNAs in their proliferation. *Proc Am Thorac Soc* 5(3):300–304.

Lu Y, Thomson JM, Wong HY, Hammond SM, Hogan BL. (2007). Transgenic over-expression of the microRNA miR-17-92 cluster promotes proliferation and inhibits differentiation of lung epithelial progenitor cells. *Dev Biol* 310(2):442–453.

Lubkin SR, Murray JD. (1995). A mechanism for early branching in lung morphogenesis. *J Math Biol* 34(1):77–94.

Lüdtke TH, Farin HF, Rudat C, *et al.* (2013). Tbx2 controls lung growth by direct repression of the cell cycle inhibitor genes Cdkn1a and Cdkn1b. *PLoS Genet* 9(1):e1003189.

Luks FI, Roggin KK, Wild YK, Piasecki GJ, Rubin LP, Lesieur-Brooks AM, *et al.* (2001). Effect of lung fluid composition on type II cellular activity after tracheal occlusion in the fetal lamb. *J Pediatr Surg* 36(1):196–201.

Maduro MF, Meneghini MD, Bowerman B, Broitman-Maduro G, Rothman JH. (2001). Restriction of mesendoderm to a single blastomere by the combined action of skn-1 and a gsk-3beta homolog is mediated by med-1 and -2 in c. Elegans. *Mol Cell* 7:475–485.

Mahler DA, Fierro-Carrion G, Baird JC. (2003). Evaluation of dyspnea in the elderly. Clin *Geriatr Med* 19(1):19–33, v.

Mahler DA, Rosiello RA, Loke J. (1986). The aging lung. *Clin Geriatr Med* 2(2):215–225.

Mailleux AA, Kelly R, Veltmaat JM, *et al.* (2005). Fgf10 expression identifies parabronchial smooth muscle cell progenitors and is required for their entry into the smooth muscle cell lineage. *Development* 132(9): 2157–2166.

Maina JN, West JB. (2005). Thin and strong! The bioengineering dilemma in the structural and functional design of the blood-gas barrier. *Physiol Rev* 85:811–844.

Marko Z, Nikolić OC, Quitz Jeng, *et al.* (2017). Human embryonic lung epithelial tips are multipotent progenitors that can be expanded *in vitro* as long-term selfrenewing organoids. Elife.

Martin-Belmonte F, Perez-Moreno M. (2011). Epithelial cell polarity, stem cells and cancer. *Nat Rev Cancer* 12(1):23–38.

Massaro GD, Massaro D. (1996). Postnatal treatment with retinoic acid increases the number of pulmonary alveoli in rats. *Am J Physiol* 270:L305–L310.

Massaro GD, Massaro D, Chambon P. (2003). Retinoic acid receptor-alpha regulates pulmonary alveolus formation in mice after, but not during, perinatal period. *Am J Physiol Lung Cell Mol Physiol* 284:L431–L433.

Matthay MA, Thompson BT, Read EJ, McKenna Jr., DH, Liu KD, Calfee CS, Lee JW. (2010). Therapeutic potential of mesenchymal stem cells for severe acute lung injury. *Chest* 138(4):965–972.

Matute-Bello G, Downey G, Moore BB, Groshong SD, Matthay MA, Slutsky AS, Kuebler, WM. Acute Lung Injury in Animals Study (2011). An official American Thoracic Society workshop report: Features and measurements of experimental acute lung injury in animals. *Am J Respir Cell Mol Biol* 44(5):725–738.

McDowell EM, Newkirk C, Coleman B. (1985). Development of hamster tracheal epithelium: II. Cell proliferation in the fetus. *Anat Rec* 213(3): 448–456.

McGee KP, Hubmayr RD, Ehman RL. (2008). MR elastography of the lung with hyperpolarized 3He. Magn. *Reson Med* 59:14–18.

McGowan S, Jackson SK, Jenkins-Moore M, Dai HH, Chambon P, Snyder JM. (2000). Mice bearing deletions of retinoic acid receptors demonstrate reduced lung elastin and alveolar numbers. *Am J Respir Cell Mol Biol* 23:162–167.

McGowan SE, Torday JS. (1997). The pulmonary lipofibroblast (lipid interstitial cell) and its contributions to alveolar development. *Annu Rev Physiol* 59:43–62.

McIntyre BA, Alev C, Mechael R, Salci KR, Lee JB, Fiebig-Comyn A, Guezguez B, Wu Y, Sheng G, Bhatia M. (2014). Expansive generation of functional airway epithelium from human embryonic stem cells. *Stem Cells Transl Med* 3(1):7–17.

Mei SH, McCarter SD, Deng Y, Parker CH, Liles WC, Stewart DJ. (2007). Prevention of LPS-induced acute lung injury in mice by mesenchymal stem cells overexpressing angiopoietin 1. *PLoS Med* 4(9):e269.

Mendelsohn C, Lohnes D, Decimo D, Lufkin T, LeMeur M, Chambon P, Mark M. (1994). Function of the retinoic acid receptors (RARs) during development (II). Multiple abnormalities at various stages of organogenesis in RAR double mutants. *Development* 120:2749–2771.

Meno C, Shimono A, Saijoh Y, Yashiro K, Mochida K, Ohishi S, Noji S, Kondoh H, Hamada H. (1998). Lefty-1 is required for left-right determination as a regulator of lefty-2 and nodal. *Cell* 94(3):287–297.

Metcalfe C, Siebel CW. (2015). The hedgehog hold on homeostasis. *Cell Stem Cell* 17(5):505–506.

Metzger RJ, Klein OD, Martin GR, *et al.* (2008). The branching programme of mouse lung development. *Nature* 453(7196):745–750.

Michos O, Panman L, Vintersten K, Beier K, Zeller R, Zuniga A. (2004). Gremlin-mediated BMP antagonism induces the epithelial-mesenchymal feedback signaling controlling metanephric kidney and limb organogenesis. *Development* 131(14):3401–3410.

Miettinen PJ. (1997). Epidermal growth factor receptor in mice and men — any applications to clinical practice? *Ann Med* 29(6):531–534.

Miettinen PJ, Berger JE, Meneses J, Phung Y, Pedersen RA, Werb Z, Derynck R. (1995). Epithelial immaturity and multiorgan failure in mice lacking epidermal growth factor receptor. *Nature* 376:337–341.

Miettinen PJ, Warburton D, Bu D, Zhao JS, Berger JE, Minoo P, Koivisto T, Allen L, Dobbs L, Werb Z, Derynck R. (1997). Impaired lung branching morphogenesis in the absence of functional EGF receptor. *Dev Biol* 186(2):224–236.

Miura T. (2015). Models of lung branching morphogenesis. *J Biochem* 157(3):121–127.

Moens CB, Auerbach AB, Conlon RA, Joyner AL, Rossant J. (1992). A targeted mutation reveals a role for N-myc in branching morphogenesis in the embryonic mouse lung. *Genes Dev* 6(5):691–704.

Moodley Y, Sturm M, Shaw K, Shimbori C, Tan DB, Kolb M, Graham R. (2016). Human mesenchymal stem cells attenuate early damage in a ventilated pig model of acute lung injury. *Stem Cell Res* 17(1):25–31.

Moore KA, Polte T, Huang S, Shi B, Alsberg E, Sunday ME, Ingber DE. (2005). Control of basement membrane remodeling and epithelial branching morphogenesis in embryonic lung by Rho and cytoskeletal tension. *Dev Dyn* 232:268–281.

Morikawa M, Derynck R, Miyazono K. (2016). TGF-β and the TGF-β family: Context-dependent roles in cell and tissue physiology. *Cold Spring Harb Perspect Biol* 8(5):pii:a021873.

Morimoto M, Liu Z, Cheng HT, Winters N, Bader D, Kopan R. (2010). Canonical Notch signaling in the developing lung is required for determination of arterial smooth muscle cells and selection of Clara versus ciliated cell fate. *J Cell Sci* 123(Pt 2):213–224.

Morrison SJ, Kimble J. (2006). Asymmetric and symmetric stem-cell divisions in development and cancer. *Nature* 441(7097):1068–1074.

Motoyama J, Liu J, Mo R, Ding Q, Post M, Hui CC. (1998). Essential function of Gli2 and Gli3 in the formation of lung, trachea and oesophagus. *Nat Genet* 20:54–57.

Mou HM, Zhao R, Sherwood R, Ahfeldt T, Lapey A, Wain J, Sicilian L, Izvolsky K, Musunuru K, Cowan C, Rajagopal J, Lau F. (2012). Generation of multipotent lung and airway progenitors from mouse ESCs and patient-specific cystic fibrosis iPSCs. *Cell Stem Cell* 10(5): 635–635.

Minoo P, Su G, Drum H, Bringas P, Kimura S. (1999). Defects in tracheo-esophageal and lung morphogenesis in Nkx2.1(–/–) mouse embryos. *Dev Biol* 209:60–71.

Moessinger AC, Harding R, Adamson TM, Singh M, Kiu GT. (1990). Role of lung fluid volume in growth and maturation of the fetal sheep lung. *J Clin Invest* 86(4):1270–1277.

Morimoto M, Liu Z, Cheng HT, Winters N, Bader D, Kopan R. (2010). Canonical Notch signaling in the developing lung is required for determination of arterial smooth muscle cells and selection of Clara versus ciliated cell fate. *J Cell Sci* 123:213–224.

Morrisey E, Cardoso WV, Lane RH, Rabinovitch M, Abman SH, Ai X, Albertine KH, Bland RD, Chapman HA, Checkley W, *et al.* (2013). Molecular determinants of lung development. *Ann Am Thorac Soc* 10:S12–S16.

Morrisey EE, Hogan BL. (2010). Preparing for the first breath: Genetic and cellular mechanisms in lung development. *Dev Cell* 18:8–23.

Morrison SJ, Kimble J. (2006). Asymmetric and symmetric stem-cell divisions in development and cancer. *Nature* 441(7097):1068–1074.

Mucenski ML, Wert SE, Nation JM, Loudy DE, Huelsken J, Birchmeier W, Morrisey EE, Whitsett JA. (2003). Beta-catenin is required for specification of proximal/distal cell fate during lung morphogenesis. *J Biol Chem* 278: 40231–40238.

Muglia LJ, Bae DS, Brown TT, Vogt SK, Alvarez JG, Sunday ME, Majzoub JA. (1999). Proliferation and differentiation defects during lung development in corticotropin-releasing hormone-deficient mice. *Am J Respir Cell Mol Biol* 20(2):181–188.

Narayanan M, Owers-Bradley J, Beardsmore CS, *et al.* (2012). Alveolarization continues during childhood and adolescence: New evidence from helium-3 magnetic resonance. *Am J Respir Crit Care Med* 185(2): 186–191.

Nelson WJ. (2003). Epithelial cell polarity from the outside looking in. *News Physiol Sci* 18:143–146.

Nirmalanandhan VS, Sittampalam GS. (2009). Stem cells in drug discovery, tissue engineering, and regenerative medicine: Emerging opportunities and challenges. *J Biomol Screen* 14(7):755–768.

Nguyen NM, Miner JH, Pierce RA, Senior RM. (2002). Laminin alpha 5 is required for lobar septation and visceral pleural basement membrane formation in the developing mouse lung. *Dev Biol* 246(2): 231–244.

Noctor SC, Martinez-Cerdeno V, Ivic L, Kriegstein AR. (2004). Cortical neurons arise in symmetric and asymmetric division zones and migrate through specific phases. *Nat Neurosci* 7:136–144.

Northway WH, Jr., Rosan RC, Porter DY. (1967). Pulmonary disease following respirator therapy of hyaline-membrane disease. Bronchopulmonary dysplasia. *N Engl J Med* 276(7):357–368.

Northway WH, Jr., RB, Moss KB, Carlisle BR, Parker RL, Popp PT. Pitlick, *et al.* (1990). Late pulmonary sequelae of bronchopulmonary dysplasia. *N Engl J Med* 323(26):1793–1799.

Nyeng P, Norgaard GA, Kobberup S, Jensen J. (2008). FGF10 maintains distal lung bud epithelium and excessive signaling leads to progenitor state arrest, distalization, and goblet cell metaplasia. *BMC Dev Biol* 8:2.

Obladen M. (2005). History of surfactant up to 1980. *Biol Neonate* 87:308–316.

O'Brodovich H.M. (1996). Immature epithelial Na+ channel expression is one of the pathogenetic mechanisms leading to human neonatal respiratory distress syndrome. *Proc Assoc Am Physicians* 108(5):345–355.

Oh SP, Li E. (1997). The signaling pathway mediated by the type IIB activin receptor controls axial patterning and lateral asymmetry in the mouse. *Genes Dev* 11(14):1812–1826.

Ohle SJ, Anandaiah A, Fabian AJ, Fine AKotton DN. (2012). Maintenance and repair of the lung endothelium does not involve contributions from marrow-derived endothelial precursor cells. *Am J Respir Cell Mol Biol* 47(1):11–19.

Okubo T, Knoepfler PS, Eisenman RN, Hogan BL. (2005). Nmyc plays an essential role during lung development as a dosage-sensitive regulator of progenitor cell proliferation and differentiation. *Development* 132(6): 1363–1374.

Oliver JR, Kushwah R, Wu J, Pan J, Cutz E, Yeger H, Waddell TK, Hu J. (2011). Elf3 plays a role in regulating bronchiolar epithelial repair kinetics following Clara cell-specific injury. *Lab Invest* 91(10):1514–1529.

Olver RE, Strang LB. (1974). Ion fluxes across the pulmonary epithelium and the secretion of lung liquid in the foetal lamb. *J Physiol* 241(2): 327–357.

Olver RE, Ramsden CA, Strang LB, Walters DV. (1986). The role of amiloride-blockable sodium transport in adrenaline-induced lung liquid reabsorption in the fetal lamb. *J Physiol* 376:321–340.

Ozkan M, Dweik RA, Ahmad M. (2001). Drug-induced lung disease. *Cleve Clin J Med* 68(9):782–785, 789–795.

Pandya HC, Innes J, Hodge R, Bustani, P, Silverman M, Kotecha S. (2006). Spontaneous contraction of pseudoglandular-stage human airspaces is associated with the presence of smooth muscle-alpha-actin and smooth musclespecific myosin heavy chain in recently differentiated fetal human airway smooth muscle. *Biol Neonate* 89:211–219.

Park WY, Miranda B, Lebeche D, Hashimoto G, Cardoso WV. (1998). FGF-10 is a chemotactic factor for distal epithelial buds during lung development. *Dev Biol* 201(2):125–134.

Parvez O, Voss AM, De Kok M, Roth-Kleiner M, Belik J. (2006). Bronchial muscle peristaltic activity in the fetal rat. *Pediatr Res* 59:756–761.

Park WS. (2016). Stem cells for the prevention of bronchopulmonary dysplasia, in V. Bhandari (ed.), *Bronchopulmonary Dysplasia, Respiratory Medicine*, Springer International Publishing, Switzerland.

Park WY, Miranda B, Lebeche D, Hashimoto G, Cardoso WV. (1998). FGF-10 is a chemotactic factor for distal epithelial buds during lung development. *Dev Biol* 201(2):125–134.

Peng T, DB, Frank RS, Kadzik MP, Morley KS, Rathi T, Wang S, Zhou L, Cheng MM, Lu EE. (2015). *Morrisey Nature* 526:578–582.

Pepicelli CV, Lewis PM, McMahon AP. (1998). Sonic hedgehog regulates branching morphogenesis in the mammalian lung. *Curr Biol* 8(19): 1083–1086.

Pera MF, Trounson AO. (2004). Human embryonic stem cells: Prospects for development. *Development* 131(22):5515–5525.

Perl AK, Wert SE, Loudy DE, Shan Z, Blair PA, Whitsett JA. (2005). Conditional recombination reveals distinct subsets of epithelial cells in trachea, bronchi, and alveoli. *Am J Respir Cell Mol Biol* 33(5): 455–462.

Peschon JJ, Slack JL, Reddy P, Stocking KL, Sunnarborg SW, Lee DC, Russell WE, Castner BJ, Johnson RS, Fitzner JN, Boyce RW, Nelson N, Kozlosky CJ, Wolfson MF, Rauch CT, Cerretti DP, Paxton RJ, March CJ, Black RA. (1998). An essential role for ectodomain shedding in mammalian development. *Science* 282(5392):1281–1284.

Pinkerton K, Harding R. (2014). *The Lung: Development, Aging and the Environment* (2nd edition). Academic Press/Elsevier, London, UK, 544pp.

Plantier L, Marchand-Adam S, Antico Arciuch VG, *et al.* (2007). Keratinocyte growth factor protects against elastase-induced pulmonary emphysema in mice. *Am J Physiol Lung Cell Mol Physiol* 293(5): L1230–L1239.

Plopper C, St George J, Cardoso W, Wu R, Pinkerton K, Buckpitt A. (1992). Development of airway epithelium. Patterns of expression for markers of differentiation. *Chest* 101(3 Suppl):2S–5S.

Polglase GR, Wallace MJ, Grant DA, Hooper SB. (2004). Influence of fetal breathing movements on pulmonary hemodynamics in fetal sheep. *Pediatr Res* 56(6):932–938.

Popova AP, Bentley JK, Anyanwu AC, *et al.* (2012). Glycogen synthase kinase-3ß/ß-catenin signaling regulates neonatal lung mesenchymal stromal cell myofibroblastic differentiation. *Am J Physiol Lung Cell Mol Physiol* 303(5):L439–8.

Powell AE, Shung C, Saylor KW, Müllendorff KA, Weiss JB, Wong MH. (2010). Lessons from development: A role for asymmetric stem cell division in cancer. *Stem Cell Res* 5(1):1–12.

Prockop DJ, Oh JY. (2012). Medical therapies with adult stem/progenitor cells (MSCs): A backward journey from dramatic results *in vivo* to the cellular and molecular explanations. *J Cell Biochem* 113(5):1460–1469.

Puissant B, Barreau C, Bourin P, Clavel C, Corre J, Bousquet C, Taureau C, Cousin B, Abbal M, Laharrague P, Penicaud L, Casteilla L, Blancher A. (2005). Immunomodulatory effect of human adipose tissue-derived adult stem cells: Comparison with bone marrow mesenchymal stem cells. *Br J Haematol* 129(1):118–129.

Quaggin SE, Schwartz L, Cui S, Igarashi P, Deimling J, Post M, Rossant J. (1999). The basic-helix-loop-helix protein pod1 is critically important for kidney and lung organogenesis. *Development* 126:5771–5783.

Que J, Luo X, Schwartz RJ, Hogan BL. (2009). Multiple roles for Sox2 in the developing and adult mouse trachea. *Development* 136:1899–1907.

Que J, Wilm B, Hasegawa H, Wang F, Bader D, Hogan BL. (2008). Mesothelium contributes to vascular smooth muscle and mesenchyme during lung development. *Proc Natl Acad Sci USA* 105(43):16626–16630.

Ramasamy SK, Mailleux AA, Gupte VV, *et al.* (2007). Fgf10 dosage is critical for the amplification of epithelial cell progenitors and for the formation of multiple mesenchymal lineages during lung development. *Dev Biol* 307(2):237–247.

Rankin SA, Gallas AL, Neto A, Gomez-Skarmeta JL, Zorn AM. (2012). Suppression of Bmp4 signaling by the zinc-finger repressors Osr1 and Osr2 is required for Wnt/beta-catenin-mediated lung specification in *Xenopus. Development* 139:3010–3020.

Rankin SA, Zorn AM. (2014). Gene regulatory networks governing lung specification. *J Cell Biochem* 115(8):1343–1350.

Rannels SR, Rannels DE. (1989). The type II pneumocyte as a model of lung cell interaction with the extracellular matrix. *J Mol Cell Cardiol* 21(Suppl 1):151–159.

Rawlins EL, Hogan BL. (2006). Epithelial stem cells of the lung: Privileged few or opportunities for many? *Development* 133(13):2455–2465.

Rawlins EL, Clark CP, Xue Y, Hogan BL. (2009). The Id2+ distal tip lung epithelium contains individual multipotent embryonic progenitor cells. *Development* 136(22):3741–3745.

Rawlins EL, Ostrowski LE, Randell SH, Hogan BL. (2007). Lung development and repair: Contribution of the ciliated lineage. *Proc Natl Acad Sci USA* 104(2):410–417.

Rawlins EL, Okubo T, Que J, Xue Y, Clark C, Luo X, Hogan BL. (2008). Epithelial stem/progenitor cells in lung postnatal growth, maintenance, and repair. *Cold Spring Harb Symp Quant Biol* 73:291–295.

Rawlins EL, Okubo T, Xue Y. (2009). The role of Scgb1a1+ Clara cells in the long-term maintenance and repair of lung airway, but not alveolar, epithelium. *Cell Stem Cell* 4(6):525–534.

Rawlins EL. (2008). Lung epithelial progenitor cells: Lessons from development. *Proc Am Thorac Soc* 5(6):675–681.

Rawlins EL, Clark CP, Xue Y, Hogan BL. (2009). The Id2+ distal tip lung epithelium contains individual multipotent embryonic progenitor cells. *Development* 136:3741–3745.

Ray P, Devaux Y, Stolz DB, *et al.* (2003). Inducible expression of keratinocyte growth factor (KGF) in mice inhibits lung epithelial cell death induced by hyperoxia. *Proc Natl Acad Sci USA* 100(10):6098–6103.

Riccardi D, Park J, Lee WS, Gamba G, Brown EM, Hebert SC. (1995). Cloning and functional expression of a rat kidney extracellular calcium/polyvalent cation-sensing receptor. *Proc Natl Acad Sci USA* 92(1):131–135.

Roberts D, Dalziel S. (2006). Antenatal corticosteroids for accelerating fetal lung maturation for women at risk of preterm birth. *Cochrane Database Syst Rev* CD004454.

Rock JR, Harfe BD. (2008). Expression of TMEM16 paralogs during murine embryogenesis. *Dev Dyn* 237(9):2566–2574.

Rock JR, Hogan BL. (2011). Epithelial progenitor cells in lung development, maintenance, repair, and disease. *Annu Rev Cell Dev Biol* 27:493–512.

Rock JR, Futtner CR, Harfe BD. (2008). The transmembrane protein TMEM16A is required for normal development of the murine trachea. *Dev Biol* 321(1):141–149.

Rock JR, Randell SH, Hogan BL. (2010). Airway basal stem cells: A perspective on their roles in epithelial homeostasis and remodeling. *Dis Model Mech* 3(9–10):545–556.

Rockich BE, Hrycaj SM, Shih HP, *et al.* (2013). Sox9 plays multiple roles in the lung epithelium during branching morphogenesis. *Proc Natl Acad Sci USA* 110(47):E4456–54.

Rodaway A, Takeda H, Koshida S, Broadbent J, Price B, Smith JC, Patient R, Holder N. (1999). Induction of the mesendoderm in the zebrafish germ ring by yolk cell-derived tgf-beta family signals and discrimination of mesoderm and endoderm by fgf. *Development* 126: 3067–3078.

Ritter MC, Jesudason R, Majumdar A, Stamenovic D, Buczek-Thomas JA, Stone PJ, Nugent MA, Suki B. (2009). A zipper network model of the failure mechanics of extracellular matrices. *Proc Natl Acad Sci USA* 106:1081–1086.

Reddy R, Buckley S, Doerken M, *et al.* (2004). Isolation of a putative progenitor subpopulation of alveolar epithelial type 2 cells. *Am J Physiol Lung Cell Mol Physiol* 286(4):L658–7.

Reynolds SD, Giangreco A, Hong KU, McGrath KE, Ortiz LA, Stripp BR. (2004). Airway injury in lung disease pathophysiology: Selective depletion of airway stem and progenitor cell pools potentiates lung inflammation and alveolar dysfunction. *Am J Physiol Lung Cell Mol Physiol* 287(6): L1256–L1265.

Reynolds SD, Giangreco A, Power JH, Stripp BR. (2000). Neuroepithelial bodies of pulmonary airways serve as a reservoir of progenitor cells capable of epithelial regeneration. *Am J Pathol* 156(1):269–278.

Rock JR, Onaitis MW, Rawlins EL, Lu Y, Clark CP, Xue Y, Randell SH, Hogan BL. (2009). Basal cells as stem cells of the mouse trachea and human airway epithelium. *Proc Natl Acad Sci USA* 106(31): 12771–12775.

Rock JR, Onaitis MW, Rawlins EL, *et al.* (2009). Basal cells as stem cells of the mouse trachea and human airway epithelium. *Proc Natl Acad Sci USA* 106(31):12771–12775.

Rojas M, Xu, J, Woods, CR, Mora, AL, Spears, W, Roman J, Brigham KL. (2005). Bone marrow-derived mesenchymal stem cells in repair of the injured lung. *Am J Respir Cell Mol Biol* 33(2):145–152.

Roszell B, Mondrinos MJ, Seaton A, Simons DM, Koutzaki SH, Fong GH, Lelkes PI, Finck CM. (2009). Efficient derivation of alveolar type II cells from embryonic stem cells for *in vivo* application. *Tiss Eng Part A* 15(11):3351–3365.

Rubin LP, Kovacs CS, De Paepe ME, Tsai SW, Torday JS, Kronenberg HM. (2004). Arrested pulmonary alveolar cytodifferentiation and defective surfactant synthesis in mice missing the gene for parathyroid hormone-related protein. *Dev Dyn* 230(2):278–289.

Rutter M, Wang J, Huang Z, Kuliszewski M, Post M. (2010). Gli2 influences proliferation in the developing lung through regulation of cyclin expression. *Am J Respir Cell Mol Biol* 42(5):615–625.

Sabapathy V, Kumar S. (2016). hiPSC-derived iMSCs: NextGen MSCs as an advanced therapeutically active cell resource for regenerative medicine. *J Cell Mol Med* 20(8):1571–1588.

Sadeghian Chaleshtori S, Dezfouli MR, Dehghan MM, Tavanaeimanesh H. (2016). Generation of lung and airway epithelial cells from embryonic stem cells *in vitro*. *Crit Rev Eukaryot Gene Expr* 26(1):1–9.

Sage EK, Loebinger MR, Polak J, Janes SM. (2008). The Role of Bone Marrow-Derived Stem Cells in Lung Regeneration and Repair, Cambridge (MA): Stem Book.

Sakiyama J, Yamagishi A, Kuroiwa A. (2003). Tbx4-Fgf10 system controls lung bud formation during chicken embryonic development. *Development* 130:1225–1234.

Sanchez-Esteban J, Cicchiello LA, Wang Y, Tsai SW, Williams LK, Torday JS, *et al.* (2001). Mechanical stretch promotes alveolar epithelial type II cell differentiation. *J Appl Physiol* 91(2):589–595.

Sanchez-Esteban J. (2013). Mechanical forces in fetal lung development: Opportunities for translational research. *Front Pediatr* 1:51.

Schira J, Falkenberg H, Hendricks M, Waldera-Lupa DM, Kogler G, Meyer HE, Muller HW, Stuhler K. (2015). Characterization of regenerative phenotype of unrestricted somatic stem cells (USSC) from human umbilical cord blood (hUCB) by functional secretome analysis. *Mol Cell Proteomics* 14(10):2630–2643.

Schittny JC. (2017). Development of the lung. *Cell a Tissue Res* 367(3): 427–444.

Schittny JC, Djonov V, Fine A, Burri PH. (1998). Programmed cell death contributes to postnatal lung development. *Am J Respir Cell Mol Biol* 18:786–793.

Schmidt M, Sun G, Stacey MA, Mori L, Mattoli S. (2003). Identification of circulating fibrocytes as precursors of bronchial myofibroblasts in asthma. *J Immunol* 171(1):380–389.

Schultz CJ, Torres E, Londos C, Torday JS. (2002). Role of adipocyte differentiation-related protein in surfactant phospholipid synthesis by type II cells. *Am J Physiol Lung Cell Mol Physiol* 283:L288–L296.

Sekine K, Ohuchi H, Fujiwara M, Yamasaki M, Yoshizawa T, Sato T, Yagishita N, Matsui D, Koga Y, Itoh N, Kato S. (1999). Fgf10 is essential for limb and lung formation. *Nat Genet* 21:138–141.

Selmani Z, Naji A, Zidi I, Favier B, Gaiffe E, Obert L, Borg C, Saas P, Tiberghien P, Rouas-Freiss N, Carosella ED, Deschaseaux F. (2008). Human leukocyte antigen-G5 secretion by human mesenchymal stem cells is required to suppress T lymphocyte and natural killer function and to induce CD4+CD25highFOXP3+ regulatory T cells. *Stem Cells* 26(1): 212–222.

Serls AE, Doherty S, Parvatiyar P, Wells JM, Deutsch GH. (2005). Different thresholds of fibroblast growth factors pattern the ventral foregut into liver and lung. *Development* 132(1):35–47.

Seymour PA, Freude KK, Tran MN, *et al.* (2007). SOX9 is required for maintenance of the pancreatic progenitor cell pool. *Proc Natl Acad Sci USA* 104(6):1865–1870.

Sgantzis N, Yiakouvaki A, Remboutsika E, Kontoyiannis DL. (2011). HuR controls lung branching morphogenesis and mesenchymal FGF networks. *Dev Biol* 354(2):267–279.

Shan L, Subramaniam M, Emanuel RL, *et al.* (2008). Centrifugal migration of mesenchymal cells in embryonic lung. *Dev Dyn* 237(3):750–757.

Shannon JM, Hyatt BA. (2004). Epithelial-mesenchymal interactions in the developing lung. *Annu Rev Physiol* 66:625–645.

Shen MM. (2007). Nodal signaling: Developmental roles and regulation. *Development* 134, 1023–1034.

Sherwood RI, Maehr R, Mazzoni EO, Melton DA. (2011). Wnt signaling specifies and patterns intestinal endoderm. *Mech Dev* 128:387–400.

Shi W, Xu J, Warburton D. (2009). Development, repair and fibrosis: What is common and why it matters. *Respirology* 14(5):656–655.

Shiels H, Li X, Schumacker PT, Maltepe E, Padrid PA, Sperling A, Thompson CB, Lindsten T. (2000). TRAF4 deficiency leads to tracheal malformation with resulting alterations in air flow to the lungs. *Am J Pathol* 157(2):679–688.

Shikama N, Lutz W, Kretzschmar R, Sauter N, Roth JF, Marino S, Wittwer J, Scheidweiler A, Eckner R. (2003). Essential function of p300 acetyltransferase activity in heart, lung and small intestine formation. *EMBO J* 22(19):5175–5185.

Shu W, Guttentag S, Wang Z, Andl T, Ballard P, Lu MM, Piccolo S, Birchmeier W, Whitsett JA, Millar SE, Morrisey EE. (2005). Wnt/beta-catenin signaling acts upstream of N-myc, BMP4, and FGF signaling to regulate proximal-distal patterning in the lung. *Dev Biol* 283(1): 226–239.

Shu W, Jiang YQ, Lu MM, Morrisey EE. (2002). Wnt7b regulates mesenchymal proliferation and vascular development in the lung. *Development* 129(20):4831–4842.

Shu W, Guttentag S, Wang Z, *et al.* (2005). Wnt/beta-catenin signaling acts upstream of N-myc, BMP4, and FGF signaling to regulate proximal-distal patterning in the lung. *Dev Biol* 283(1):226–239.

Shu W, Lu MM, Zhang Y, Tucker PW, Zhou D, Morrisey EE. (2007). Foxp2 and Foxp1 cooperatively regulate lung and esophagus development. *Development* 134(10):1991–2000.

Sinclair SE, Molthen RC, Haworth ST, Dawson CA, Waters CM. (2007). Airway strain during mechanical ventilation in an intact animal model. *Am J Respir Crit Care Med* 176:786–794.

Smith VC, Zupancic JAF, McCormick MC, Croen, LA, Greene J, Escobar GJ, *et al.* (2005). Trends in severe bronchopulmonary dysplasia rates between 1994 and 2002. *J Pediatr* 146(4):469–447.

Sock E, Rettig SD, Enderich J, Bösl MR, Tamm ER, Wegner M. (2004). Gene targeting reveals a widespread role for the high-mobility-group transcription factor Sox11 in tissue remodeling. *Mol Cell Biol* 24(15): 6635–6644.

Sountoulidis A, Stavropoulos A, Giaglis S, *et al.* (2012). Activation of the canonical bone morphogenetic protein (BMP) pathway during lung morphogenesis and adult lung tissue repair. *PLoS One* 7(8):e41460.

Spaeth, JM, Hunter, CS, Bonatakis L, Guo M, French CA, Slack I, Hara M, Fisher SE, Ferrer J, Morrisey EE, Stanger BZ, Stein R. (2015). The FOXP1, FOXP2 and FOXP4 transcription factors are required for islet alpha cell proliferation and function in mice. *Diabetologia* 58: 1836–1844.

Spaggiari GM, Capobianco A, Becchetti S, Mingari MC, Moretta L. (2006). Mesenchymal stem cell-natural killer cell interactions: Evidence that activated NK cells are capable of killing MSCs, whereas MSCs can inhibit IL-2-induced NK-cell proliferation. *Blood* 107(4):1484–1490.

Spees JL, Olson SD, Ylostalo J, Lynch PJ, Smith J, Perry A, Peister A, Wang MY, Prockop DJ. (2003). Differentiation, cell fusion, and nuclear fusion during *ex vivo* repair of epithelium by human adult stem cells from bone marrow stroma. *Proc Natl Acad Sci USA* 100(5):2397–2402.

Spurlin JW, Nelson CM. (2017). Building branched tissue structures: From single cell guidance to coordinated construction. *Philos Trans R Soc Lond B Biol Sci* 372(1720).

Srinivasan S, Strange J, Awonusonu F, Bruce MC. (2002). Insulin-like growth factor I receptor is downregulated after alveolarization in an apoptotic fibroblast subset. *Am J Physiol Lung Cell Mol Physiol* 282:L457–L467.

Stabler CT, Morrisey EE. (2017). Developmental pathways in lung regeneration. *Cell Tissue Res* 367(3):677–685.

Steele-Perkins G, Plachez C, Butz KG, Yang G, Bachurski CJ, Kinsman SL, Litwack ED, Richards LJ, Gronostajski RM. (2005). The transcription factor gene Nfib is essential for both lung maturation and brain development. *Mol Cell Biol* 25(2):685–698.

Sterner-Kock A, Thorey IS, Koli K, Wempe F, Otte J, Bangsow T, Kuhlmeier K, Kirchner T, Jin S, Keski-Oja J, von Melchner H. (2002). Disruption of the gene encoding the latent transforming growth factor-beta binding protein 4 (LTBP-4) causes abnormal lung development, cardiomyopathy, and colorectal cancer. *Genes Dev* 16: 2264–2273.

Stevens T, Phan S, Frid MG, Alvarez D, Herzog E, Stenmark KR. (2008). Lung vascular cell heterogeneity: Endothelium, smooth muscle, and fibroblasts. *Proc Am Thorac Soc* 5(7):783–791.

Suen HC, Bloch KD, Donahoe PK. (1994). Antenatal glucocorticoid corrects pulmonary immaturity in experimentally induced congenital diaphragmatic hernia in rats. *Pediatr Res* 35:523–529.

Sugahara K, Iyama KI, Kimura T, Sano K, Darlington GJ, Akiba T, Takiguchi M. (2001). Mice lacking CCAAt/enhancer-binding protein-alpha show hyperproliferation of alveolar type II cells and increased surfactant protein mRNAs. *Cell Tissue Res* 306(1):57–63.

Suki B, Bates JH. (2008). Extracellular matrix mechanics in lung parenchymal diseases. *Respir Physiol Neurobiol* 163:33–43.

Sun J, Chen H, Chen C, Whitsett JA, Mishina Y, Bringas P Jr, Ma JC, Warburton D, Shi W. (2008). Prenatal lung epithelial cell-specific abrogation of Alk3-bone morphogenetic protein signaling causes neonatal respiratory distress by disrupting distal airway formation. *Am J Pathol* 172(3):571–582.

Suratt BT, Cool CD, Serls AE, Chen L, Varella-Garcia M, Shpall EJ, Brown KK, Worthen GS. (2003). Human pulmonary chimerism after hematopoietic stem cell transplantation. *Am J Respir Crit Care Med* 168(3):318–322.

Takahashi K, Mitsui K, Yamanaka S. (2003). Role of ERas in promoting tumour-like properties in mouse embryonic stem cells. *Nature* 423(6939):541–545.

Takahashi Y, Izumi Y, Kohno M, *et al.* (2010). Thyroid transcription factor-1 influences the early phase of compensatory lung growth in adult mice. *Am J Respir Crit Care Med* 181(12):1397–406.

Tan Y, AlKhamees B, Jia D, Li L, Couture JF, Figeys D, Jinushi M, Wang L. (2015). MFG-E8 Is critical for embryonic stem cell-mediated T cell immunomodulation. *Stem Cell Reports* 5(5):741–752.

Tapscott SJ. (2005). The circuitry of a master switch: Myod and the regulation of skeletal muscle gene transcription. *Development* 132(12): 2685–2695.

Tarran R, Button B, Picher M, Paradiso AM, Ribeiro CM, Lazarowski ER, Zhang L, Collins PL, Pickles RJ, Fredberg JJ, Boucher RC. (2005). Normal and cystic fibrosis airway surface liquid homeostasis. The effects of phasic shear stress and viral infections. *J Biol Chem* 280: 35751–35759.

Tarran R, Button B, Boucher RC. (2006). Regulation of normal and cystic fibrosis airway surface liquid volume by phasic shear stress. *Annu Rev Physiol* 68:543–561.

Tarran R, Donaldson S, Boucher RC. (2007). Rationale for hypertonic saline therapy for cystic fibrosis lung disease. *Semin Respir Crit Care Med* 28:295–302.

Tata PR, Rajagopal J. (2017). Plasticity in the lung: Making and breaking cell identity. *Development* 144(5):755–766.

Tawhai MH, Burrowes KS. (2008). Multi-scale models of the lung airways and vascular system. *Adv Exp Med Biol* 605:190–194.

Tefft D, De Langhe SP, Del Moral PM, et al. (2005). A novel function for the protein tyrosine phosphatase Shp2 during lung branching morphogenesis. *Dev Biol* 282(2):422–431.

Tefft D, Lee M, Smith S, Crowe DL, Bellusci S, Warburton D. (2002). mSprouty2 inhibits FGF10-activated MAP kinase by differentially binding to upstream target proteins. *Am J Physiol Lung Cell Mol Physiol* 283(4):L700–L706.

Tepass U. (2012). The apical polarity protein network in Drosophila epithelial cells: Regulation of polarity, junctions, morphogenesis, cell growth, and survival. *Annu Rev Cell Dev Biol* 28:655–685.

Theise ND, Henegariu O, Grove J, Jagirdar J, Kao PN, Crawford JM, Badve S, Saxena R, Krause DS. (2002). Radiation pneumonitis in mice: A severe injury model for pneumocyte engraftment from bone marrow. *Exp Hematol* 30(11):1333–1338.

Thibeault DW, Truog WE, Ekekezie II. (2003a). Acinar arterial changes with chronic lung disease of prematurity in the surfactant era. *Pediatr Pulmonol* 36(6):482–429.

Thibeault DW, Mabry SM, Ekekezie II, Zhang X, Truog WE. (2003b). Collagen scaffolding during development and its deformation with chronic lung disease. *Pediatrics* 111(4 Pt 1):766–776.

Thurlbeck WM. (1983a). Postpneumonectomy compensatory lung growth. *Am Rev Respir Dis* 128(6):965–967.

Thurlbeck WM. (1983b). Overview of the pathology of pulmonary emphysema in the human. *Clin Chest Med* 4(3):337–350.

Tompkins DH, Besnard V, Lange AW, Keiser AR, Wert SE, Bruno MD, Whitsett JA. (2011). Sox2 activates cell proliferation and differentiation in the respiratory epithelium. *Am J Respir Cell Mol Biol* 45(1):101–110.

Toonkel RL, Hare JM, Matthay MA, Glassberg MK. (2013). Mesenchymal stem cells and idiopathic pulmonary fibrosis. Potential for clinical testing. *Am J Respir Crit Care Med* 188(2):133–140.

Torday JS, Rehan VK. (2002). Stretch-stimulated surfactant synthesis is coordinated by the paracrine actions of PTHrP and leptin. *Am J Physiol Lung Cell Mol Physiol* 283(1):L130–L135.

Torday JS, Torday DP, Gutnick J, Qin J, Rehan V. (2001). Biologic role of fetal lung fibroblast triglycerides as antioxidants. *Pediatr Res* 49(6): 843–849.

Torday JS, Sun H, Wang L, Torres E, Sunday ME, Rubin LP. (2002). Leptin mediates the parathyroid hormone-related protein paracrine stimulation of fetal lung maturation. *Am J Physiol Lung Cell Mol Physiol* 282(3): L405–L410.

Tropea KA, Leder E, Aslam M, Lau AN, Raiser DM, Lee JH, Balasubramaniam V, Fredenburgh LE, Alex Mitsialis S, Kourembanas S, Kim CF. (2012). Bronchioalveolar stem cells increase after mesenchymal stromal cell treatment in a mouse model of bronchopulmonary dysplasia. *Am J Physiol Lung Cell Mol Physiol* 302(9):L829–L837.

Tropea KA, Leder E, Aslam M, Lau AN, Raiser DM, Lee J, *et al.* (2012). Bronchioalveolar stem cells increase after mesenchymal stromal cell treatment in a mouse model of bronchopulmonary dysplasia. *Am J Physiol Lung Cell Mol Physiol* 302(9):L829–L837.

Tweedell KS. (2017). The adaptability of somatic stem cells: A review. *J Stem Cells Regen Med* 13(1):3–13.

Tyson JE, Wright LL, Oh W, Kennedy KA, Mele L, Ehrenkranz RA, *et al.* (1999). Vitamin A supplementation for extremely-low-birth-weight infants. National Institute of Child Health and Human Development Neonatal Research Network. N *Engl J Med* 340:1962–1968.

Uehara M, Yashiro K, Takaoka K, Yamamoto M, Hamada H. (2009). Removal of maternal retinoic acid by embryonic cyp26 is required for correct nodal expression during early embryonic patterning. *Genes Dev* 23:1689–1698.

Unbekandt M, del Moral PM, Sala FG, Bellusci S, Warburton D, Fleury V. (2008). Tracheal occlusion increases the rate of epithelial branching of embryonic mouse lung via the FGF10-FGFR2b-Sprouty2 pathway. *Mech Dev* 125:314–324.

Usui H, Shibayama M, Ohbayashi N, Konishi M, Takada S, Itoh N. (2004). Fgf18 is required for embryonic lung alveolar development. *Biochem Biophys Res Commun* 322(3):887–892.

van Haaften T, Byrne R, Bonnet S, Rochefort GY, Akabutu J, Bouchentouf M, et al. (2009). Airway delivery of mesenchymal stem cells prevents arrested alveolar growth in neonatal lung injury in rats. *Am J Respir Crit Care Med* 180(11):1131–1142.

Vaughan AE, Brumwell AN, Xi Y, Gotts JE, Brownfield DG, Treutlein B, Tan K, Tan V, Liu FC, Looney MR, Matthay MA, Rock JR, Chapman HA. (2015). Lineage-negative progenitors mobilize to regenerate lung epithelium after major injury. *Nature* 517(7536):621–625.

Vergnes L, Péterfy M, Bergo MO, Young SG, Reue K. (2004). Lamin B1 is required for mouse development and nuclear integrity. *Proc Natl Acad Sci USA* 101(28):10428–10433.

Wagers AJ, Weissman IL. (2004). Plasticity of adult stem cells. *Cell* 116(5): 639–648.

Walsh JR, Chambers DC, Davis RJ, Morris NR, Seale HE, Yerkovich ST, Hopkins PM. (2013a). Impaired exercise capacity after lung transplantation is related to delayed recovery of muscle strength. *Clin Transplant* 27(4):E504–E511.

Walsh JR, McKeough ZJ, Morris NR, Chang AT, Yerkovich ST, Seale HE, Paratz JD. (2013b). Metabolic disease and participant age are independent predictors of response to pulmonary rehabilitation. *J Cardiopulm Rehabil Prev* 33(4):249–256.

Walsh MC, Szefler S, Davis J, Allen M, van Marter L, Abman S, et al. (2006). Summary proceedings from the bronchopulmonary dysplasia group. *Pediatrics* 117(3 Pt 2):S52–S56.

Wan H, Dingle S, Xu Y, Besnard V, Kaestner KH, Ang SL, Wert S, Stahlman MT, Whitsett JA. (2005). Compensatory roles of Foxa1 and Foxa2 during lung morphogenesis. *J Biol Chem* 280(14):13809–13816.

Wang C, Chang KC, Somers G, et al. (2009). Protein phosphatase 2A regulates self-renewal of Drosophila neural stem cells. *Development* 136(13): 2287–2296.

Wang D, Haviland DL, Burns AR, Zsigmond E, Wetsel RA. (2007). A pure population of lung alveolar epithelial type II cells derived from human embryonic stem cells. *Proc Natl Acad Sci USA* 104(11):4449–4454.

Wang IC, Zhang Y, Snyder J, Sutherland MJ, Burhans MS, Shannon JM, Park HJ, Whitsett JA, Kalinichenko VV. (2010). Increased expression of FoxM1 transcription factor in respiratory epithelium inhibits lung sacculation and causes Clara cell hyperplasia. *Dev Biol* 347(2): 301–314.

Wang J, Edeen K, Manzer R, Chang Y, Wang S, Chen X, Funk CJ, Cosgrove GP, Fang X, Mason RJ. (2007). Differentiated human alveolar epithelial cells and reversibility of their phenotype *in vitro*. *Am J Respir Cell Mol Biol* 36(6):661–668.

Wang JH, Deimling S, D'Alessandro NE, Zhao L, Possmayer F, Drysdale TA. (2011). Retinoic acid is a key regulatory switch determining the difference between lung and thyroid fates in Xenopus laevis. *BMC Dev Biol* 11:75.

Wang X, Wang Y, Snitow M, Stewart K, Li S, Lu M, Morrisey EE. (2016). Expression of histone deacetylase 3 instructs alveolar type I cell differentiation by regulating a Wnt signaling niche in the lung. *Dev. Biol.* 414(2):161–169.

Wang Y, Maciejewski BS, Soto-Reyes D, Lee HS, Warburton D, Sanchez-Esteban J. (2009). Mechanical stretch promotes fetal type II epithelial cell differentiation via shedding of HB-EGF and TGF-alpha. *J Physiol.* 587(Pt 8):1739–1753, doi:10.1113/jphysiol.2008.163899.

Wang Y, Tian Y, Morley MP, Lu MM, Demayo FJ, Olson EN, Morrisey EE. (2013). Development and regeneration of Sox2þ endoderm progenitors are regulated by a Hdac1/2-Bmp4/Rb1 regulatory pathway. *Dev Cell* 24:345–358.

Wani MA, Wert SE, Lingrel JB. (1999). Lung Kruppel-like factor, a zinc finger transcription factor, is essential for normal lung development. *J Biol Chem* 274(30):21180–21185.

Wansleeben C, Barkauskas CE, Rock JR, Hogan BL. (2013). Stem cells of the adult lung: Their development and role in homeostasis, regeneration, and disease. *Wiley Interdiscip Rev Dev Biol* 2(1):131–148.

Warburton D, El-Hashash AHK, Carraro G, *et al.* (2010). Lung organogenesis. *Curr Top Dev Biol* 90:73–158.

Warburton D, Olver B. (1997). Coordination of genetic, epigenetic, and environmental factors in lung development, injury, and repair. *Chest* 111:119S–122S.

Warburton D, Perin L, Defilippo R, Bellusci S, Shi W, Driscoll B. (2008). Stem/progenitor cells in lung development, injury repair, and regeneration. *Proc Am Thorac Soc* 5:703–706.

Warburton D, Schwarz M, Tefft D, Flores-Delgado G, Anderson KD, Cardoso WV. (2000). The molecular basis of lung morphogenesis. *Mech Dev* 92(1):55–81.

Warburton D, Seth R, Shum L, Horcher PG, Hall FL, Werb Z, Slavkin HC. (1992). Epigenetic role of epidermal growth factor expression and signalling in embryonic mouse lung morphogenesis. *Dev Biol* 149:123–133.

Warburton D, Tefft D, Mailleux A, Bellusci S, Thiery JP, Zhao J, Buckley S, Shi W, Driscoll B. (2001). Do lung remodeling, repair, and regeneration recapitulate respiratory ontogeny? *Am J Respir Crit Care Med* 164(10 Pt 2):S59–S62.

Ward HE, Nicholas TE. (1984). Alveolar type I and type II cells. *Aust N Z J Med* 14(5 Suppl 3):731–734.

Warga RM, Nusslein-Volhard C. (1999). Origin and development of the zebrafish endoderm. *Development* 126:827–838.

Warner BB, LA, Stuart RA, Papes JR. (1998). Wispe functional and pathological effects of prolonged hyperoxia in neonatal mice. *Am J Physiol* 275(1 Pt 1):L110–L117.

Weaver M, Batts L, Hogan BL. (2003). Tissue interactions pattern the mesenchyme of the embryonic mouse lung. *Dev Biol* 258:169–184.

Weaver M, Yingling JM, Dunn NR, Bellusci S, Hogan BL. (1999). Bmp signaling regulates proximal-distal differentiation of endoderm in mouse lung development. *Development* 126:4005–4015.

Weibel EB. (1984). The Pathway for Oxygen: Structure and Function in the Mammalian Respiratory System, Harvard University Press, Cambridge, Mass, 1984.

Weinstein M, Xu X, Ohyama K, Deng CX. (1998). FGFR-3 and FGFR-4 function cooperatively to direct alveogenesis in the murine lung. *Development* 125(18):3615–3623.

Weiss DJ. (2014). Current status of stem cells and regenerative medicine in lung biology and diseases. *Stem Cells* 32(1):16–25.

Weiss DJ, Bertoncello I, Borok Z, Kim C, Panoskaltsis-Mortari A, Reynolds S, Rojas M, Stripp B, Warburton D, Prockop DJ. (2011).

Stem cells and cell therapies in lung biology and lung diseases. *Proc Am Thorac Soc* 8(3):223–272.

Weiss DJ, Casaburi R, Flannery R, LeRoux-Williams M, Tashkin DP. (2013). A placebo-controlled, randomized trial of mesenchymal stem cells in COPD. *Chest* 143(6):1590–1598.

Weiss DJ, Ortiz LA. (2013). Cell therapy trials for lung diseases: Progress and cautions. *Am J Respir Crit Care Med* 188(2):123–125.

Wendel DP, Taylor DG, Albertine KH, Keating MT, Li DY. (2000). Impaired distal airway development in mice lacking elastin. *Am J Respir Cell Mol Biol* 23:320–326.

Whitsett JA, Weaver TE. (2002). Hydrophobic surfactant proteins in lung function and disease. *N Engl J Med* 347:2141–2148.

Whitsett JA, Wert SE, Trapnell BC. (2004). Genetic disorders influencing lung formation and function at birth. *Hum Mol Genet* 13 Spec No 2: R207–R215.

Wilson JM, DiFiore JW, Peters CA. (1993). Experimental fetal tracheal ligation prevents the pulmonary hypoplasia associated with fetal nephrectomy: Possible application for congenital diaphragmatic hernia. *J Pediatric Surge* 28:1433–1440.

Wilson-Costello D, Friedman H, Minich N, Fanaroff AA, Hack M. (2005). Improved survival rates with increased neurodevelopmental disability for extremely low birth weight infants in the 1990s. *Pediatrics* 115(4): 997–1003.

Wodarz A. (2002). Establishing cell polarity in development. *Nat Cell Biol* 4(2):E39–E44.

Wong AP, Bear CE, Chin S, Pasceri P, Thompson TO, Huan LJ, Ratjen F, Ellis J, Rossant J. (2012). Directed differentiation of human pluripotent stem cells into mature airway epithelia expressing functional CFTR protein. *Nat Biotechnol* 30(9):876–U108.

Woods DF, Wu JW, Bryant PJ. (1997). Localization of proteins to the apico-lateral junctions of Drosophila epithelia. *Dev Genet* 20(2): 111–118.

Wu JY, Feng L, Park HT, Havlioglu N, Wen L, Tang H, Bacon KB, Jiang Z, Zhang X, Rao Y. (2001). The neuronal repellent Slit inhibits leukocyte chemotaxis induced by chemotactic factors. *Nature* 410:948–952.

Xian J, Clark KJ, Fordham R, Pannell R, Rabbitts TH, Rabbitts PH. (2001). Inadequate lung development and bronchial hyperplasia in mice with a targeted deletion in the Dutt1/Robo1 gene. *Proc Natl Acad Sci USA* 98:15062–15066.

Xu J, Qu J, Cao L, Sai Y, Chen C, He L, Yu L. (2008). Mesenchymal stem cell-based angiopoietin-1 gene therapy for acute lung injury induced by lipopolysaccharide in mice. *J Pathol* 214(4):472–481.

Xu K, Nieuwenhuis E, Cohen BL, Wang W, Canty AJ, Danska JS, Coultas L, Rossant J, Wu MY, Piscione TD, Nagy A, Gossler A, Hicks GG, Hui CC, Henkelman RM, Yu LX, Sled JG, Gridley T, Egan SE. (2010). Lunatic Fringe-mediated Notch signaling is required for lung alveogenesis. *Am J Physiol Lung Cell Mol Physiol* 298(1):L45–L56.

Xu P, Yu S, Jiang R, Kang C, Wang G, Jiang H, Pu P. (2009). Differential expression of Notch family members in astrocytomas and medulloblastomas. *Pathol Oncol Res* 15(4):703–710.

Yalcin HC, Perry SF, Ghadiali SN. (2007). Influence of airway diameter and cell confluence on epithelial cell injury in an *in vitro* model of airway reopening. *J Appl Physiol* 103:1796–1807.

Yamashita YM. (2009). The centrosome and asymmetric cell division. *Prion* 3:84–88.

Yamashita YM, Yuan H, Cheng J, Hunt AJ. (2010). Polarity in stem cell division: Asymmetric stem cell division in tissue homeostasis. *Cold Spring Harb Perspect Biol* 2:a001313.

Yan C, Sever Z, Whitsett JA. (1995). Upstream enhancer activity in the human surfactant protein B gene is mediated by thyroid transcription factor 1. *J Biol Chem* 270:24852–24857.

Yan B, Omar FM, Das K, Ng WH, Lim C, Shiuan K, et al. (2008). Characterization of Numb expression in astrocytomas. *Neuropathology* 28:479–484.

Yang CE, Jiang J, Yang X, Wang H, Du J. (2016). Stem/progenitor cells in endogenous repairing responses: New toolbox for the treatment of acute lung injury. *J Transl Medi* 14:47.

Yang H, Lu MM, Zhang L, Whitsett JA, Morrisey EE. (2002). GATA6 regulates differentiation of distal lung epithelium. *Development* 129(9): 2233–2246.

Yang L, Naltner A, Yan C. (2003). Overexpression of dominant negative retinoic acid receptor alpha causes alveolar abnormality in transgenic neonatal lungs. *Endocrinology* 144:3004–3011.

Yang S, Ma K, Geng Z, Sun X, Fu X. (2015). Oriented cell division: New roles in guiding skin wound repair and regeneration. *Biosci Reports* 35(6):e00280.

Yao L, Liu CJ, Luo Q, Gong M, Chen J, Wang LJ, Huang Y, Jiang X, Xu F, Li TY, Shu C. (2013). Protection against hyperoxia-induced lung fibrosis by KGF-induced MSCs mobilization in neonatal rats. *Pediatr Transplant* 17(7):676–682.

Yeh TF, Lin YJ, Lin HC, Huang CC, Hsieh WS, Lin CH, *et al.* (2004). Outcomes at school age after postnatal dexamethasone therapy for lung disease of prematurity. *N Engl J Med* 350(13):1304–1313.

Yin Y, Wang F, Ornitz DM. (2011). Mesothelial- and epithelial-derived FGF9 have distinct functions in the regulation of lung development. *Development* 138 (15):3169–3177.

Zaret KS. (2016). From endoderm to liver bud: Paradigms of cell type specification and tissue morphogenesis. *Curr Top Dev Biol* 117:647–669.

Zemke AC, Teisanu RM, Giangreco A, Drake JA, Brockway BL, Reynolds SD, Stripp BR. (2009). beta-Catenin is not necessary for maintenance or repair of the bronchiolar epithelium. *Am J Respir Cell Mol Biol* 41(5):535–543.

Zhao J, Chen H, Peschon JJ, Shi W, Zhang Y, Frank SJ, Warburton D. (2001). Pulmonary hypoplasia in mice lacking tumor necrosis factor-alpha converting enzyme indicates an indispensable role for cell surface protein shedding during embryonic lung branching morphogenesis. *Dev Biol* 232:204–218.

Zhen G, Liu H, Gu N, Zhang H, Xu Y, Zhang Z. (2008). Mesenchymal stem cells transplantation protects against rat pulmonary emphysema. *Front Biosci* 13:3415–3422.

Zhou Q, Law AC, Rajagopal J, Anderson WJ, Gray PA, Melton DA. (2007). A multipotent progenitor domain guides pancreatic organogenesis. *Dev Cell* 13(1):103–114.

Zhou, Z, You Z. (2016). Mesenchymal stem cells alleviate LPS-induced acute lung injury in mice by MiR-142a-5p-controlled pulmonary endothelial cell autophagy. *Cell Physiol Biochem* 38(1):258–266.

Zhu J, Wen W, Zheng Z, *et al.* (2011). Structures of the LGN/mInsc and LGN/NuMA complexes suggest distinct functions of the Par3/mInsc/LGN and Gαi/LGN/NuMA pathways in asymmetric cell division. *Molecular Cell* 43(3):418–431.

Zhu Y, Chen X, Yang X, El-Hashash A. (2018). Stem cells in lung repair and regeneration: Current applications and future promise. *J Cell Physiol* 1–11.

Zimdahl B, Ito T, Blevins A, Bajaj J, Konuma T, Weeks J, Koechlein CS, Kwon HY, Arami O, Rizzieri D, Broome HE, Chuah C, Oehler VG, Sasik R, Hardiman G, Reya T. (2014). Lis1 regulates asymmetric division in hematopoietic stem cells and in leukemia. *Nat Genet* 46(3): 245–252.

Zorn AM, Wells JM. (2007). Molecular basis of vertebrate endoderm development. *Int Rev Cytol* 259:49–111.

Zorn AM, Wells JM. (2009). Vertebrate endoderm development and organ formation. *Annu Rev Cell Dev Biol* 25:221–251.

Index

www.ingramcontent.com/pod-product-compliance
Lightning Source LLC
Chambersburg PA
CBHW050627190326
41458CB00008B/2165